Visions of Electric Media

Visions of Electric Media

Television in the Victorian and Machine Ages

Ivy Roberts

Routledge
Taylor & Francis Group
LONDON AND NEW YORK

First published in 2019 by Amsterdam University Press Ltd.

Published 2025 by Routledge
4 Park Square, Milton Park, Abingdon, Oxon OX14 4RN
605 Third Avenue, New York, NY 10158

Routledge is an imprint of the Taylor & Francis Group, an informa business

ISBN: 9789462986596 (hbk)
ISBN: 9781041190325 (pbk)
ISBN: 9781003709152 (ebk)
NUR 670

Cover illustration: 'Professor Goaheadison's Latest,' *Fun*, 3 July 1889, 6.

Cover design: Coördesign, Leiden

DOI: 10.5117/9789462986596

Table of Contents

List of Figures

Introduction: The Lifespan of a Media Technology

Abstract

How do 'new' media become old? What determines the 'newness' of a technology? This introductory chapter engages questions related to history, technology, culture, and communication as well as examines the cultural context in which a new technology informs our understanding of its changing identity. The success and failure of new technologies depends heavily on the cultural expectations of its emergence. This study of early 'speculative era' television begins with a foundation in the history and theory of new media and a call to adopt an historically reflective view of technological development. After laying a foundation for the study of 'new' media and technological change, this chapter provides an overview of the chapters and what to expect.

Keywords: Media history; media theory; cultural studies; media archaeology; new media; television

In its most literal form, 'tele-vision' means 'seeing at a distance' (*tele*=distant, *videre*=to see). Television relates to the uses of technological and electrical apparatuses that make it possible to see into the distance. That is, the viewer is not looking out over the horizon with bare eyes, or even gazing through a telescope. The precise term, 'television', was coined in 1900.[1] Before then, television was known by many names, such as the telectroscope, the telephonoscope, and the diaphote. Nineteenth-century inventors dreamt of 'seeing by electricity'. After 1900, new names emerged, indicating a growing cultural obsession with television: the telephote, the Ikonophone, the

1 Constantin Perskyi, 'Television Using Electricity,' in *Congres International D. Electricite (International Congress on Electricity of 1900)* (Paris: Gauthier –Villars, 1901); R.W. Burns, *Television: An International History of the Formative Years* (London: Institutions of Electrical Engineers, 1998), 106; George Shiers, *Early Television: A Bibliographic Guide* (London: Taylor & Francis, 1997), 36.

Roberts, I., *Visions of Electric Media: Television in the Victorian and Machine Ages.*
Taylor & Francis Group, 2019
DOI 10.5117/9789462986596_INTRO

Ikonoscope. The prospect of visual radio (radiovision) became popular, but people wondered whether it would be a fleeting fad. By the mid twentieth century, television became synonymous with broadcast and programming. By the time television emerged as a mass media in the 1950s, it had become *a fixed idea.*

Over the course of what has been termed its 'speculative era', 'television' evolved from a rough concept into a functional technology, eventually becoming a mass medium.[2] Television's speculative era spans two distinct periods. The Victorian Age, broadly understood as the reign of Queen Victoria of England (1837-1901), more generally applies to late nineteenth-century culture. The Machine Age, a term coined by historian of technology Lewis Mumford, refers to the early 20th-century culture of scientific progress, manufacturing, and industry.[3]

The first printed reference to anything resembling an electronic screen appeared in 1877.[4] On the heels of Alexander Graham Bell's invention and demonstrations of the telephone, speculation began to circulate about a visual adjunct that would make it possible to see the person on the other end of the line. In announcements of new inventions and breakthroughs, journalists bemoaned the death of theatre.[5] Satirists illustrated caricatures of the apocalyptic consequences of new media.[6] Rumours circulated about how the very foundation of society would change irrevocably upon the arrival of a device for 'seeing by electricity'. Electronic capabilities to support

2 R.W. Burns, 'Part I: The Era of Speculation 1877 to c. 1922,' in *Television: An International History of the Formative Years* (London: Institutions of Electrical Engineers, 1998), 3-140; Shiers, *Early Television*; Andreas Fickers, 'Television,' in *The Handbook of Communication History*, ed. Peter Simonson (New York: Routledge, 2013), 239.

3 Lewis Mumford, *Technics and Civilization* (Chicago, IL: University of Chicago Press, 2010), 9; John M. Jordan, *Machine-Age Ideology: Social Engineering and American Liberalism, 1911-1939* (Chapel Hill, NC: University of North Carolina Press, 1994); David F. Noble, *America By Design: Science, Technology, and the Rise of Corporate Capitalism* (Cambridge, UK: Oxford University Press, 1979); J. P. Telotte, *A Distant Technology: Science Fiction Film and the Machine Age* (Middletown, CT: Wesleyan University Press, 1999); David A. Mindell, *Between Human and Machine: Feedback, Control, and Computing Before Cybernetics* (Baltimore, MD: Johns Hopkins University Press, 2002), 1-2; Nicholas Daly, 'The Machine Age,' in *The Oxford Handbook of Modernisms*, ed. Peter Osborne, Peter Brooker, Andrzej Gasiorek, Deborah Longworth, and A. J. Thacker (Cambridge, UK: Oxford University Press, 2010).

4 The Electrician, 'The Electroscope,' letter to the editor, *New York Sun*, 30 March 1877; Louis Figuier, 'Le Telectroscope,' in *L'Annee Scientifique et Industrielle* (1877).

5 'Edison's Last,' *Boston Journal*, 13 May 1889; 'What Edison Claims,' *Chicago Journal*, 13 May 1891.

6 'Untitled' (This discovery mania), *New York Tribune*, 14 May 1878; 'Professor Goaheadison's Latest,' *Fun*, 3 July 1889.

a television industry on a practical basis showed signs of maturity by the year 1930. In the intervening 50 years or so, a rich, diverse visual culture of 'television' emerged, creating expectations and myths about new media.

There is no question that, today, television is again undergoing a transition.[7] Some call it the 'death of television'. Others herald the birth of convergence culture.[8] The shift from broadcast to online delivery indicates a change in platform as well as the form of television programming. With such changes, one might ask if watching YouTube or Netflix constitutes a television experience, or if a new term is required altogether. Given the uncertain state of television today, there is no time like the present to ask, 'when was television born'? To understand what television has become in the early 21st century, we need to look back to what television was before it became a mass medium.

Like the death of print, talk has been circulating for over a decade about the death of television. Television superseded radio, and the web and mobile media will supersede television, they say. Television producers and studios bemoan falling Nielsen ratings, declining viewership, and leaks and bootlegs. Fears such as these should be countered by recognizing that the media exist in harmony with culture. The media evolve, responding to cultural needs, economic capabilities, and technological possibilities. We need a new paradigm; we need to replace 'supersession' with 'transition'.

This book engages the question of how new media and technology come into being, and the forces that conspire to bring one possible future into existence over another. Televisual culture in the Victorian and Machine ages created expectations about what television would become: idealistic, maybe; futuristic, certainly. 'Television' coalesced decades before engineers made images appear on screens. Engineers and inventors, journalists and writers, and artists and philosophers contributed to a dense and varied discussion about 'television'. The general public also played an important role in constructing expectations about new technology. Responses to media announcements, letters to the editor, and rumour-laden op-eds document the vast range of speculations, anxieties, and expectations that existed

7 Amanda Lotz, *The Television Will be Revolutionized,* second edition (New York: NYU Press, 2014); Jean Burgess and Joshua Green, *YouTube: Online Video and Participatory Culture* (John Wiley & Sons, 2013); Peele Snickars and Patrick Vonderau, eds, *The YouTube Reader* (Stockholm: National Library of Sweden, 2009).

8 c.f. William Uricchio, 'The Future of a Medium Once Known as Television,' in *The Youtube Reader*, eds. Peele Snickars and Patrick Vonderau (Stockholm: National Library of Sweden, 2009), 24; Henry Jenkins, 'What Happened Before Youtube?' in *YouTube: Online Video and Participatory Culture* by Jean Burgess and Joshua Green (John Wiley & Sons, 2013), 109.

before the rise of mainstream popular science. Television's speculative era constitutes a distinct televisual culture in which agents (audiences, producers, inventors, engineers, critics, etc.) contribute to ways of thinking about 'seeing by electricity'. The visual rhetoric and discourse of television's speculative era provides a baseline for the study of media in transition.

This book promotes a conception of technology as a product of the imagination, sometimes called 'imaginary media'.[9] The possibilities of technology are limited only to what the inventor can dream and what the engineer can make possible. From an historical perspective, there is little difference between the imaginings of a science-fiction author and the inventor's actual functional technology. The visual rhetoric and discourse of television's speculative era highlights the peculiar similarities between science fiction and invention. Science-fiction authors wrote about it. Albert Robida 'invented' FaceTime, in the form of the telephonoscope, in 1882.[10] Hugo Gernsback prophesized modern uses of television in his Telephote (1918).[11] Inventors toiled over designs and filed patents. Journalists speculated on the cultural uses of television. Science-fiction texts, along with news articles, rumours, and satirical illustrations, converge into a discourse that creates expectations about new technologies and media.

A New Approach to Old Media

The early 21st-century media landscape has been labelled a convergence culture and a participatory culture. Futurist thinking identifies 'emerging media' as if some progress were taking place. But 'new media' is a misnomer. The very term clouds the fact that media are in a state of constant transition. Are we living in a post-cinema or post-television age? In order to address the question of whether or not a change is really taking place, we need to rephrase the question itself. If we move past media form and supersession, the conversation becomes one of historical change and media in transition.

9 Eric Kluitenberg, 'On the Archaeology of Imaginary Media,' in *Media Archaeology: Approaches, Applications, and Implications*, eds. Erkki Huhtamo and Jussi Parikka (Berkeley, CA: University of California Press, 2011), 48-69; Simone Natale and Gabriele Balbi, 'Media and the Imaginary in History: The Role of the Fantastic in Different Stages of Media Change,' *Media History* 20, no. 2 (2014): 203-218.

10 Albert Robida, *Le Vingtième Siècle* (*The Twentieth Century*), (Paris: Librairie Illustree, 1882; Middletown, CT: Wesleyan University Press, 2004).

11 Hugo Gernsback, 'Television and the Telephot,' *Electrical Experimenter*, May 1918.

A contextual approach to media history acknowledges that every 'old' or obsolete technology was once new, novel, and innovative. Media historians who adopt this approach have examined early photography, telegraphy, and motion pictures in many contexts. In their volume *New Media, Old Media*, editors Wendy Chun and Thomas Keenan suggest an historical approach to 'new media' or 'emerging media'.[12] David Thornburg and Henry Jenkins prefer the designation 'media in transition'.[13] Film historian Rick Altman calls his method 'crisis historiography', referring to media in historical periods of identity crisis and change. 'We find that the technology today confidently called cinema was for over a decade considered quite differently by its contemporaries. In their early years, projected moving images were subjected to multiple contradictory definitions and treatments[...]. New technologies are always born nameless.'[14] In *Silent Film Sound*, Altman reveals that the very name we use today to refer to cinema before the coming of sound establishes an historically contingent designation that gets in the way of understanding how contemporary audiences would have experienced it.

'Media in transition' acknowledges that technologies are always in a state of flux. This is certainly true today, as the meaning of television and film are adapting to new systems for production and delivery. In a post-broadcast age, media industries are struggling with ways to identify film ('digital cinema'?) and television ('digital video content'?). Perhaps instead of focussing on the form, we can redirect our attention to the way these new configurations change and adapt to social and cultural conditions. Just because Netflix makes it easier for viewers to binge on television programmes does not mean that it was impossible to do so before digital content delivery. The habit became more pronounced, but it is not an entirely new behaviour.

Because media technologies are in a process of constant flux, reacting and adjusting to cultural and technological conditions, I adopt an interdisciplinary framework that accommodates a variety of perspectives. I approach production, practice, representation, and reception as various forms of discourse. Along the way, I acknowledge historical context and the many uses of media. Like media in transition, interdisciplinarity is always unsettled. Never satisfied with a single interpretive lens, interdisciplinarity aims to

12 Wendy Chun, 'Introduction: Did Somebody Say New Media?', in *New Media, Old Media: A History and Theory Reader*, eds. Wendy Chun and Thomas Keenan (New York: Routledge, 2006), 1-2. See also Benjamin Peters, 'And Lead Us Not Into Thinking the New is New: a Bibliographic Case for New Media History,' *New Media & Society* 11, no. 1-2 (2009): 13-30.

13 David Thornburn and Henry Jenkins, eds. *Rethinking Media Change: The Aesthetics of Transition* (Cambridge, MA: MIT Press, 2003).

14 Rick Altman, *Silent Film Sound* (New York: Columbia University Press, 2007), 19.

account for the multidimensionality and complexity of its objects of study. This book conceptualizes the products and events of late nineteenth- and early 20th-century cultural history, the history of technology and science, and media history as examples of an ongoing *visual rhetoric and discourse* of 'tele-vision'.

The visual rhetoric and discourse of speculative-era televisual culture comprises:

- Satirical illustrations published in magazines and newspapers;
- Illustrations of inventions published in technical journals and popular literature;
- Perspectives of journalists published in newspapers, magazines, and books, which comment upon and document the negative, positive, and conflicted views about this new invention of 'seeing by electricity';
- Commentary published in newspaper and magazine editorials and letters to the editor;
- So-called 'ego-documents' published in newspapers, magazines, and books by inventors and their supporters.

In identifying these resources as a *discourse*, this book draws on Foucauldian-inspired cultural history.[15] French philosopher and historian Michel Foucault developed two methods for the study of history throughout his career: archaeology and genealogy. I use a mixture of both, as they have been incorporated into current methods in the study of cultural history.

Whereas Foucault proposes archaeology as a method for the 'history of the present', his genealogy has been adopted more earnestly. But historians generally consider archaeology to be deterministic, a way of telling the story of the past teleologically. In comparison to archaeology, genealogy conveys a greater appreciation for the similarities and differences between historical periods, for cultural forces and agencies.

Additionally, art historians have adapted the method of semiotics for the study of images.[16] From this perspective, images as well as words constitute a discourse: communicating and contributing to an ongoing conversation.

15 Michel Foucault, *The Archaeology of Knowledge* (New York: Pantheon, 1972); Aletta Biersack and Lynn Avery Hunt, *The New Cultural History* (Berkeley, CA: University of California Press, 1989); Victoria Bonnell, Lynn Avery Hunt, and Richard Biernacki, *Beyond the Cultural Turn: New Directions in the Study of Society and Culture* (Berkeley, CA: University of California Press, 1999); John Neubauer, ed. *Cultural History after Foucault* (Piscataway, NJ: Transaction Publishers, 1999).

16 Lynda Nead, *Myths of Sexuality: Representations of Women in Victorian Britain* (Oxford, UK: Blackwell, 1988). See also Roland Barthes, *Image-Music-Text* (New York, 1977); Michael Baxandall, *Patterns of intention: On the historical explanation of pictures* (Yale University Press,

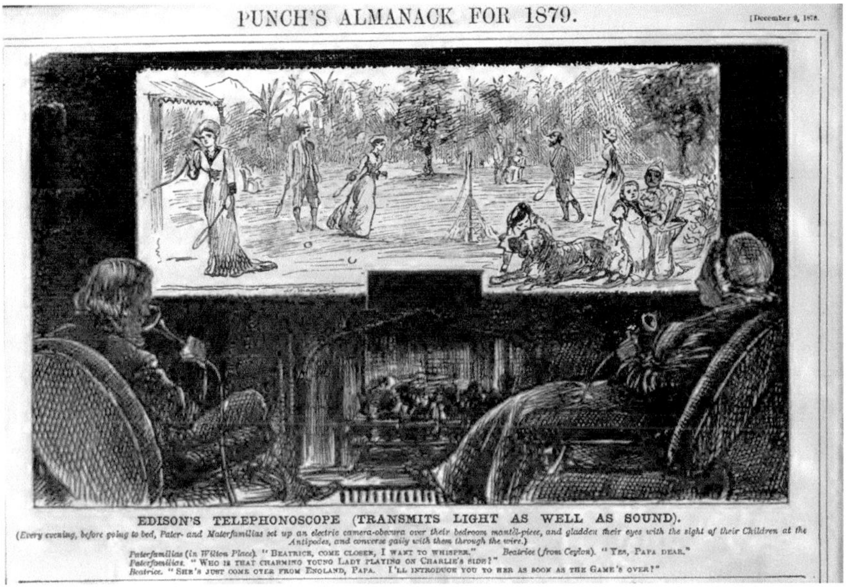

Figure 1. George Du Maurier, 'Edison's Telephonoscope,' Punch, 9 December 1878.

Notably, Lynda Hunt's *Myths of Sexuality* extends a method for the semiotic reading of images in art and history. The emerging fields of visual studies, visual culture studies, and visual studies in science address the visual rhetoric and discourse of images in culture, science, and history.[17]

The Birth of a New Medium

In 1878, *Punch* illustrator and humorist George du Maurier drew a picture of 'Edison's Telephonoscope' (Figure 1). Imagining it as a marvelous new invention that could connect two remote places, the satire attributed the telephonoscope to American inventor Thomas Edison. Edison had recently made a name for himself with the phonograph and the carbon telephone. In 1878, Edison also unveiled several new devices that seemed to push the limits of what was possible. The megaphone was said to allow the deaf to

1985); Mieke Bal and Norman Bryson, 'Semiotics and Art History,' *The Art Bulletin* 73, No. 2 (June 1991), 174-208.

17 James Elkins, *Visual Studies: A Skeptical Introduction* (New York: Routledge, 2003); W. J. T. Mitchell, *What Do Pictures Want?: The Lives and Loves of Images* (Chicago, IL: University of Chicago Press, 2005); Oliver Grau and Thomas Veigl, *Imagery in the 21st Century* (Cambridge, MA: MIT Press, 2011).

hear. The electric light would revolutionize power and energy, bringing a source of clean and safe illumination to the home. Du Maurier's illustration suggested the next new thing: visual telephony. The telephonoscope would supersede the telephone before it even reached the market.

While Du Maurier's illustration looks to us like a television screen, a nineteenth-century observer would not have seen it that way. With no concept of 'television', a reader of *Punch* might have noticed a resemblance between the telephonoscope and a mirror or an enormous photograph-come-to-life. Alexander Graham Bell had only recently begun demonstrating his *talking telegraph* to a popular audience. It would be decades before the telephone reached that audience. This period of discovery mania sometimes called the technological revolution sparked discussion in scientific communities about the possibility of extending the range of vision as well as hearing. The sky was the limit. Popular rumour also contributed to these discussions, fuelling speculation and helping to construct expectations about new technology.

A contemporary reader would have drawn associations between Du Maurier's 'Telephonoscope' and current rumours about new technologies. They might have recognized that Edison had in fact announced the invention of a telephonoscope to the public earlier that year. Since the initial neologism *telephonoscope* attracted so much ridicule, Edison settled on calling his 'ear telescope' a megaphone (Figure 2).[18] A contemporary reader could not have failed to recognize in Du Maurier's 'Telephonoscope' a hint of satire and commentary on Edison's current enterprise, electric light. In October 1878, Edison's announcement of the invention of electric light sparked furious debate and speculation. Discussion ranged from exaggerated promotion to denial and rejection. *Punch*'s December 1878 issue reflected and contributed to the ongoing discussion of new technology and its effects on everyday life.

To a 21st-century observer, however, 'Edison's Telephonoscope' is nothing but a television. The goal of media history scholarship should be to address the discrepancy between these two perspectives. In this book, I propose to examine speculative-era moving-image media technologies with the goal of uncovering trends, shifts, and continuities. By acknowledging historical, contextual, technological, and cultural perspectives, media-history scholars can become aware of the many uses of old technology. Such a perspective reconfigures the 'Telephonoscope' from a television into a satire of electric light.

18 'Edison's 'Ear Telescope,' *New York Sun*, 8 June 1878.

Figure 2a. Thomas Alva Edison, Sketch of the Telephonoscope, Laboratory Notebooks, 2 April 1878.

Figure 2b. James E. Kelly, Sketch of the Telephonoscope, Laboratory Notebooks, 17 May 1878.

Chapter Outline

By defining 'tele-vision' as 'seeing at a distance' (*tele*=distant, *videre*=to see), *Visions of Electric Media* adopts a view of media technology in a constant state of transition. Any given cultural moment assigns an identity to a media form that supports, encourages, and facilitates the needs of audiences and inventors alike. Media identity is intertwined with its culture of origin. By looking at the late nineteenth- and early 20th-century expectations of television, it is the goal of this book to defamiliarize the concept of television (presumably an immutable form), in order to speculate on the cultural uses, connections, and intersections of media in transition. Each chapter examines the relation of media technologies to conceptions of the body, the evolution of systems and networks, relationships between humans and machines, and enduring themes of control, communication, and ocularcentrism.

Visions of Electric Media investigates the many fluxing identities of 'television' before the invention of a functional apparatus in the late 1920s. The book consists of five chapters, organized into two parts. Part One examines late nineteenth-century speculative-era television, focussing on the cultures and technologies of 'seeing by electricity'. Part Two investigates the 20th-century transitions from mechanical to electronic engineering, focussing on the work performed at Bell Laboratories. By looking at the evolving discourse of television across the Victorian and Machine ages, I identify continuities and discontinuities between televisual culture in each age.

The telephonoscope emerged in the context of late nineteenth-century 'discovery mania'. Victorian engineers began the project of designing television as a means to 'annihilate space'. They pictured mirror-like screens and devices modelled after electrical telegraphs. The faith in the inevitability of technological progress drove inventors and audiences alike. Even though 'seeing by electricity' went down in history as a pipe dream, it continued to provide the impetus for engineers to develop televisual systems into the next century.

The assumptions underlying 'seeing by electricity' shaped the expectations and reception of television in later periods. A major transformation occurred at the turn of the century when engineers began to adopt a new way of thinking about distance communications. Engineers discarded the simple notion that a standalone device could do the work of bridging a gap between two places. While the analogy between the mechanical screen and the human eye persisted, designs matured into the reality of large technical systems. Machine-Age engineers adopted a new way of thinking about television that encompassed the wide expanse between viewers at either end of the circuit as part of a technological system.

When engineers finally achieved success with television systems in the 1920s, it became apparent that electronic screens were not going to live up to the Victorian ideal of the magic mirror. These technological systems constructed a new relationship between viewer and screen, displacing the Victorian expectation of the 'annihilation of space' with a modern communications paradigm. The cognitive dissonance between the knowledge of a person on the other end of the line and the picture one saw on the screen was a challenging perception to overcome.

The Machine-Age press documented these first impressions, from newspaper announcements and magazine editorials to lectures, textbooks, and popular science. But while journalists and readers continued to imagine 'seeing by electricity' as a sign of the 'annihilation of space', engineers struggled to explain the processes in which electrical signals travelled along telephone cables and over radio waves. Engineers took on the job of making the presence of the person depicted on the screen seem as real as possible. While they recognized that there was nothing natural about such a relationship, their goal became to make the act of using television as effortless as possible. Television's speculative era marks the shift from a culture of face-to-face presence to one of simulated presence. The modern, designed world grew up into artificial spaces engineered to seem natural.

If it has not already become apparent, the interdisciplinary study of speculative-era television history involves a diverse library of literature and methodology. As such, I have introduced the foundational works that support a contextual media history informed by cultural history and the history of technology. No single method accounts for both the technological development and the cultural construction of speculative-era television history. Therefore, I take advantage of different methods in each chapter.

Chapters One and Two investigate late nineteenth-century cultures and inventions for 'seeing by electricity': telephonoscopes, telectroscopes, telegraphic photography, and the Kinetograph and Kinetoscope. Inventors, engineers, and journalists alike identified 'seeing by electricity' as a tool for connecting people across vast distances, principally a form of visual telephony. I begin by investigating the circumstances surrounding George Du Maurier's illustration of 'Edison's Telephonoscope' published in *Punch* in December 1878. Media coverage on both sides of the Atlantic met Edison's inventions with a mixture of zeal and scepticism. Contemporary readers would have been acutely aware of Edison's invention of the telephonoscope (ear telescope or megaphone) and his claims to the invention of electric light that fall. I introduce the concept of 'technological folklore' to account for the rumours, hearsay, and journalist commentary that contributed to the construction of cultural representations of the telephonoscope and electric light.[19] Building on works from print history such as Lisa Gitelman's *Scripts and Grooves*, Altick's *Punch*, and Secord's *Victorian Sensation*, I argue that 'Edison's Telephonoscope' represents 'discovery mania' by negotiating between the exaggerated claims of invention and the satirical rejection of new technology for its own sake.[20] This chapter encourages media historians to weigh the presentist perspective, which associates 'Edison's Telephonoscope' with a television or electronic screen, with the view of the contemporary reader, who would have made sense of the depiction not as a prophecy, but as a speculation and critique of technology.

Chapter Two leaps ahead ten years in order to address the aftermath. In 1889, Edison announced his invention of a 'Far-Sight Machine' in the lead-up to the Columbian Exposition (1893). When he unveiled his Kinetograph and Kinetoscope to the public in 1891, the subsequent confusion fuelled speculation of a hybrid electric-photography instrument that could transmit live images (like a television) as well as reproduce scenes (like the cinema). This discussion dovetailed into early cinema mythology that bolstered its identity as a spectacular attraction.

Film historians place this moment within the context of early cinema: the demonstration of the Kinetograph and the initial press surrounding the success of creating the first filmstrips like *Fred Ott's Sneeze* and the

19 Lisa Gitelman, *Scripts, Grooves and Writing Machines: Representing technology in the Edison era* (Stanford, CA: Stanford University Press, 1999); Trevor Blank, ed., *Folk Culture in the Digital Age: The Emergent Dynamics of Human Interaction* (Boulder, CO: University Press of Colorado, 2012); Mary Jo Hatch and Michael Owen Jones, 'Photocopylore at Work: Aesthetics, Collective Creativity and the Social Construction of Organizations,' *Studies in Cultures, Organizations and Societies* 3, no. 2 (1997): 263-287; Robert Darnton, 'The Symbolic Element in History,' *The Journal of Modern History* 58, no. 1 (1986): 218-234; Richard Dorson, ed., *Folklore and Folklife* (Chicago, IL: University of Chicago Press, 1972); Lois Monteiro, 'The Electronic Pocket Calculator: Joke 1,' *Western Folklore* 35, no. 1 (1976): 75; Michael Preston, 'Xerox-lore,' *Keystone Folklore* 19, no. 1 (1974): 11-26; Michael Preston, 'Traditional Humor from the Fax Machine: All of a Kind,' *Western Folklore* 53, no. 2 (1994): 147-169.

20 Altick, *Punch*; Gerry Beegan, *The Mass Image: A Social History of Photomechanical Reproduction in Victorian London* (London: Palgrave Macmillan UK, 2008); Bernard Carlson and Michael E. Gorman, 'Understanding Invention as a Cognitive Process: The Case of Thomas Edison and Early Motion Pictures, 1888-91,' *Social Studies of Science* 20, no. 3 (1990): 387-430; Theresa Collins, Lisa Gitelman, and Gregory Jankunis, *Thomas Edison and Modern America: A Brief History with Documents* (New York: Palgrave Macmillan, 2002); Leary, *The Punch Brotherhood*; Lightman, *Victorian Popularizers of Science;* Mussel, *Science, Time and Space*; Noakes, '*Punch* and Comic Journalism'; Noakes, 'Science in Mid-Victorian *Punch*'; Secord, *Victorian Sensation;* Thomas, *Pictorial Victorians;* Andrekos Varnava, 'Punch and the British Occupation of Cyprus in 1878,' *Byzantine and Modern Greek Studies* 29, no. 2 (2005): 167-186.

Serpentine Dance.[21] Looking at the reactions to Edison's announcement in the popular press reveals a different picture. Journalists attacked Edison's proposed invention as a potential surveillance device, expressing the fears that the machine would make it possible for men to watch women undressing, for husbands cheat on their wives, and for strangers to eavesdrop on private conversations. The critical public rejection of the 'Far-sight machine' compounded fears of social and technological change. The press pictured Edison's 'Far-sight machine' as the missing link between the telephonoscope and the Kinetograph. It would be able to transmit living scenes like a visual telegraph as well as project the dramas of an opera performance. The 'Far-Sight Machine' destabilizes assumptions about the distinctions between cinema and television, transmission and recording. It shows how the identities of cinema and television are deeply entangled with the social circumstances of their emergence. By focussing on the way the 'Far-sight machine' transformed into the Kinetograph, this chapter emphasizes the distinction between the electric telescope and the production of recorded moving pictures.

Chapter Three investigates the emergence of systems thinking in the historical development of television around the turn of the 20th century. By placing the two periods side by side, this chapter fills in the gaps between the Victorian conception of 'seeing by electricity' and the Machine-Age construction of electronic screens. The scientific developments that facilitated electronic technology and the sociopolitical philosophy of efficiency contributed to a new conception of television. I examine the systems approach that emerged in engineering and the associated philosophy of technology that came with it. While the rhetoric of the annihilation of space that had propelled nineteenth-century progress never completely went away, it was displaced by a belief that human beings should adapt to the new, artificial environments made possible by the giant leaps forward in science and technology. This chapter builds on the history and philosophy of technology, including works by Mitcham, Morus, Hughes, Stielger, Hansen,

21 Tom Gunning, 'The Cinema of Attraction,' *Wide Angle* 3, no. 4 (1986); Tom Gunning, 'An Aesthetic of Astonishment'; Gordon Hendricks, *The Edison Motion Picture Myth* (Berkeley, CA: University of California Press, 1961); Gordon Hendricks, *The Kinetoscope: America's First Commercially Successful Motion Picture Exhibitor* (New York: Beginnings of the American Film, 1966); Charles Musser, *Before the Nickelodeon: Edwin S. Porter and the Edison Manufacturing Company* (Berkeley, CA: University of California Press, 1991); Paul Spehr, *The Man Who Made Movies: W.K.L. Dickson* (New Barnet, UK: John Libbey Publishing, 2008).

and Haraway.[22] Comparing and contrasting extension theory and systems thinking shows how this new philosophy of technology contributed to a new way of thinking about 'distant electric vision'.[23]

Part Two examines the changes that television underwent in the Machine Age. Chapter Four seeks to uncover the practices and ideas that drove the development of mechanical television. Since most histories of television consider electronic versions to be the first, earlier mechanical-optical systems are largely ignored.[24] From within the context of Machine-Age culture, however, radio, telephone, and movie industries vied for control over the new market for television, each with their own conception of what the new technology might become. Bell Laboratories advocated their mechanical Ikonophone; RCA backed the all-electronic system; General Electric worked to engineer a hybrid model called tele-cinema.

22 Carl Mitcham, *Thinking Through Technology: The Path Between Engineering and Philosophy* (Chicago, IL: University of Chicago Press, 1994); Iwan Rhys Morus, '"The Nervous System of Britain": Space, Time and the Electric Telegraph in the Victorian Age,' *The British Journal for the History of Science* 33, no. 4 (2000): 455-475; Bernard Stiegler, *Technics and Time: The Fault of Epimetheus* (Stanford, CA: Stanford University Press, 1998); Mark B.N. Hansen, 'Media Theory,' *Theory, Culture & Society* 23, no. 2-3 (2006): 297-306; Donna Haraway, *Simians, Cyborgs and Women: The Reinvention of Nature* (New York: Routledge, 1991); Philip Brey, 'Technology as Extension of Human Faculties,' in *Metaphysics, Epistemology and Technology,* ed. Carl Mitcham (London: Elsevier/JAI Press, 2000), 59-78.

23 John Mingers, *Realising Systems Thinking: Knowledge and Action in Management Science* (New York: Springer, 2006); Peter Checkland, *Systems Thinking, Systems Practices* (New York: John Wiley, 1981); Darrel Arnold, ed. *Traditions of Systems Theory: Major Figures and Contemporary Developments* (New York: Routledge, 2014); Thomas Hughes, *Networks of power: electrification in Western society, 1880-1930* (Baltimore, MD: Johns Hopkins University Press, 1993); Thomas Hughes, 'The Evolution of Large Technological Systems,' in *The Social Construction of Technological Systems: New Directions in the Sociology and History of Technology,* eds. Trevor Pinch, Thomas Hughes, and Wiebe Bijker (Cambridge, MA: MIT Press, 1987); Erik Van der Vleuten, 'Large Technical Systems,' in *A Companion to the Philosophy of Technology,* eds. Jan Olsen, Stig Andur Pedersen, and Vincent F. Hendricks (Malden, MA: John Wiley & Sons, 2012), 218-222.

24 Albert Abramson, *The History of Television, 1880-1941* (Jefferson, NC: McFarland & Company, 1987); R.W. Burns, 'The Contributions of the Bell Telephone Laboratories to the Early Development of Television,' *History of Technology* 13 (1991): 181-213; Charles Rubinstein, 'Optics at Bell Laboratories – General Optics, Television, and Vision,' *Applied Optics* 11, no. 11 (1972): 2401-2411; George Shiers, 'The Rise of Mechanical Television, 1901–1930,' *SMPTE Journal* 90, no. 6 (1981): 508-521; Jan Van den Ende, Wim Ravesteijn, and Dirk De Wit, 'Shaping the early Development of Television,' *Technology and Society Magazine, IEEE* 16, no. 4 (1997): 13-26; Antonio Perez Yuste, 'La Televisión Mecánica' (The Mechanical Television), in *Detrás de la Cámara. Historia de la Televisión y de sus Cincuenta Años en España* (*Behind the Scenes: History of Television and its Fifty Years in Spain*) (Madrid: Colegio Oficial de Ingenieros de Telecomunicación, 2008), 65-82; Tina Zeise, *Geschichte und Technik des Analogen Fernsehens* (*History and Technology of Analogue Television*) (Munich: GRIN Verlag, 2006).

The future of television was decided in a battle over formats and standards. Though the all-electronic models eventually won out, the outcome was far from clear in the 1920s. I aim to reveal a situation in which mechanical television can be understood as an achievement and an innovation rather than a dead-end technology. The genesis of mechanical television was closely tied with advancements in electrical engineering under the rubric of what came to be known as illuminating engineering. Little has been written about the new practice of illuminating engineering that emerged in the early 20th century.[25]

Alongside Machine-Age fields of human engineering and scientific management, illuminating engineering brought together practitioners with expertise in electrical engineering, design, and the psychology of vision.[26] This intensely interdisciplinary art and practice trained engineers in designing and implementing interior lighting environments to make them seem effortless and natural. Illuminating engineers adopted the Machine-Age philosophy of efficiency, struggling with standards and definitions that would establish a foundation for thinking about how the human, sometimes called a 'human seeing-machine', would adapt to life under electric light.[27] Photoelectric cells were the key component for making mechanical television work. These cells were popularly known as electric eyes, and they took on a symbolic function in demonstrating the correspondence between the 'human seeing-machine' and the television system. I aim to show how a Machine-Age conception of the human-machine relationship developed according to a new vision of dynamic systems, and how a pervasive attitude of control and efficiency governed and guided the further development of television technology. As such, I rely on Jordan and Noble's political history of the Machine Age and

25 Chris Otter, *The Victorian Eye: A Political History of Light and Vision in Britain, 1800-1910* (Chicago, IL: University of Chicago Press, 2008); Sean Johnston, *A History of Light and Colour Measurement: Science in the Shadows* (Bristol, UK: Institute of Physics Publishing, 2001).

26 Jennifer Alexander, 'Efficiencies of Balance Technical Efficiency, Popular Efficiency, and Arbitrary Standards in the Late Progressive Era USA,' *Social Studies of Science* 38, no. 3 (2008): 323-349; Jennifer Alexander, *The Mantra of Efficiency: From Waterwheel to Social Control* (Baltimore, MD: Johns Hopkins University Press, 2008); Sharon Corwin, 'Picturing Efficiency: Precisionism, Scientific Management, and the Effacement of Labor,' *Representations* 84, no. 1 (2003): 139-165; Graeme Gooday, *The Morals of Measurement: Accuracy, Irony, and Trust in Late Victorian Electrical Practice* (Cambridge, UK: Cambridge University Press, 2004); Toma Sorin-George, Ana-Maria Grigore, and Paul Marinescu, 'The Emergence of Scientific Management in America,' *Manager* 19 (2014): 128-131; Kenneth Thompson, introduction to *Scientific Management*, by Frederick Taylor (New York: Routledge, 2003).

27 Matthew Luckiesh, 'The Human Seeing-Machine,' *Transactions of the Illuminating Engineering Society* 27 (1932): 699-722.

histories of technology that focus on the evolution of cybernetics and control theory, including works by Beniger, Mindell, Mayer, Rabinbach, and Hayles.[28]

In stark contrast to the cathode ray tube that became the standard component in electronic television screens, mechanical television engineering took stock of the visual perception of brightness and colour. Illuminating engineers constructed an image of the average observer, which served as the standard on which all models for interior electric lighting and television were measured. Chapter Four explores the genesis of illuminating engineering in the early 20th century, its role in the efficiency movement, and the consequences it entails for the conception of the average viewer.

Chapter Five presents a case study of Bell Labs's two-way television project, also called the 'Ikonophone'. I explore archival documents charting the system's development, the role illuminating engineering played in the Ikonophone's design, the way engineers and Bell spokespeople explained the machine in newspaper and magazine media, and the reception based on eyewitness user accounts. Looking closely at press coverage of the Ikonophone project and the language journalists, writers of popular science, and engineers used to describe their experiences interacting with the screen, this chapter examines the reception and construction of television in the 1920s. Engineers described the method of designing an environment in which the users were made to feel as if they were face-to-face with the distant party. Witnesses responded to what they saw on the screen with a confused combination of metaphors, mingling expressions of a feeling of closeness with reactions to its uncanny artificiality. In stark contrast to the expectations of seeing by electricity, which supported the sense that television would provide a kind of window, distant electric vision introduced the screen as a representation of the real thing. Once the viewer beheld the screen and experienced firsthand the distinction between the onscreen and far-off presence, great transformations occurred in the conception of

28 John Jordan, *Machine-Age Ideology: Social Engineering and American Liberalism, 1911-1939* (Chapel Hill, NC: University of North Carolina Press, 1994); David Noble, *America by Design: Science, Technology, and the Rise of Corporate Capitalism* (Cambridge, UK: Oxford University Press, 1979); James Beniger, *The Control Revolution: Technological and Economic Origins of the Information Society* (Cambridge, MA: Harvard University Press, 2009); David Mindell, *Between Human and Machine: Feedback, Control, and Computing Before Cybernetics* (Baltimore, MD: Johns Hopkins University Press, 2002); Anson Rabinbach, *The Human Motor: Energy, Fatigue, and the Origins of Modernity* (Berkeley, CA: University of California Press, 1992); N. Katherine Hayles, *How We Became Posthuman: Virtual Bodies in Cybernetics, Literature, and Informatics* (Chicago, IL: University of Chicago Press, 1999).

television as a technology as well as a visual medium. The Ikonophone marks the moment of television's transition from a technology into a visual medium.

This final chapter aims to draw parallels and identify changes that occurred since the Edison age of the telephonoscope. For all intents and purposes, the Ikonophone was conceived in the image of the telephonoscope. Engineers and journalists consistently appealed to the Victorian prophecy of 'seeing by electricity'. But, as it had been reimagined through the Machine-Age conception of 'distant electric vision', the Ikonophone grew into something quite different from what anyone could have expected. Television takes on a different shape once it becomes possible to see pictures flickering on a screen.

Continuities and discontinuities are immediately perceptible in the areas of communications and media. The popular American fascination with engineering and electrical technology mirrors the late nineteenth-century discovery manias. Where satirical magazines and newspaper journalists were mouthpieces for late nineteenth-century technological folklore, Machine-Age trade publications gave popular science a new twist. While American culture sustained an idealistic image for the amateur inventor, by the early 20th century, the broadcasting and cinema industries gained a stranglehold over the direction of new technology. This chapter also draws connections between Machine-Age engineering practices and broader 20th-century approaches to screen-mediated communication and design, including the field known as presence research and studies in digital subjectivity.[29]

Throughout the five chapters of this study, I foreground issues of cultural and technological change in the interest of moving past the limitations of discourse specific to individual disciplines. More than a study in the history of television, it might be just as appropriate to describe this book as an exploration of interdisciplinary methodologies for the study of 'media in transition'. The study of 'media in transition' aims to uncover the meaning of cultural representations in context and analyse trends in the development of new technologies. Instead of focussing on a single invention, I endeavour to locate conceptions of 'television', broadly defined. In what circumstances did the idea emerge? Which factors contributed to its popularity? And how did social groups adapt to new relationships established by technological

29 Sherry Turkle, *Alone Together: Why We Expect More from Technology and Less from Each Other* (New York: Basic Books, 2011); Nancy Baym, *Personal Connections in the Digital Age* (Malden, MA: Polity, 2010); Mark Poster, 'The Digital Subject and Cultural Theory,' in *What's the Matter with the Internet* (Minneapolis, MN: University of Minnesota Press, 2001); N. Katherine Hayles, 'Virtual Bodies and Flickering Signifiers,' *October* 66 (1993): 69-91.

systems? In asking questions such as these, both continuities and discontinuities become apparent. Between the 1880s and the 1920s, social groups reacted with a mixture of adulation and scepticism to forces of technological progress. The role of satire and commentary in print media both reflected and contributed to expectations about new technology.

Moreover, the importance of face-to-face interaction in the nineteenth century gave way to a 20th-century construction of a modern viewer. Analysing this shift, as represented in cultural representations of television — satirical illustrations, technical diagrams, newspaper announcements, and popular science — brings to light changing historical attitudes concerning communication. While it may not be possible to discover the value placed on face-to-face interaction in previous eras, the transition to an age of screen-mediated communication bears relevance to an historical perception of the changing relationships between humans and nature.

Bibliography

Abramson, Albert. *The History of Television, 1880-1941.* Jefferson, NC: McFarland & Company, 1987.

Alexander, Jennifer. 'Efficiencies of Balance Technical Efficiency, Popular Efficiency, and Arbitrary Standards in the Late Progressive Era USA.' *Social Studies of Science* 38, no. 3 (2008): 323-349.

Alexander, Jennifer. *The Mantra of Efficiency: From Waterwheel to Social Control.* Baltimore, MD: Johns Hopkins University Press, 2008.

Altick, Richard. *Punch: The Lively Youth of a British Institution, 1841-1851.* Columbus: Ohio State University Press, 1997.

Altman, Rick. *Silent Film Sound.* New York: Columbia University Press, 2007.

Arnold, Darrell, ed. *Traditions of Systems Theory: Major Figures and Contemporary Developments.* New York: Routledge, 2014.

Bal, Mieke and Norman Bryson. 'Semiotics and Art History.' *The Art Bulletin* 73, no. 2 (Jun 1991): 174-208.

Barthes, Roland. *Image-Music-Text.* London: Fontana Press, 1977.

Baxandall, Michael. *Patterns of intention: On the historical explanation of pictures.* New Haven: Yale University Press, 1985.

Baym, Nancy. *Personal Connections in the Digital Age.* Malden, MA: Polity, 2010.

Beegan, Gerry. *The Mass Image: A Social History of Photomechanical Reproduction in Victorian London.* London: Palgrave Macmillan UK, 2008.

Beniger, James. *The Control Revolution: Technological and Economic Origins of the Information Society.* Cambridge, MA: Harvard University Press, 2009.

Biersack, Aletta, and Lynn Avery Hunt. *The New Cultural History.* Berkeley: University of California Press, 1989.

Bijker Wiebe, Thomas Hughes, and Trevor Pinch, eds. *The Social Construction of Technological Systems: New Directions in The Sociology and History of Technology.* Cambridge, MA: MIT Press, 1987.

Blank, Trevor, ed. *Folk Culture in the Digital Age: The Emergent Dynamics of Human Interaction.* Boulder: University Press of Colorado, 2012.

Bonnell, Victoria E., Lynn Avery Hunt, and Richard Biernacki. *Beyond the Cultural Turn: New Directions in the Study of Society and Culture.* Berkeley: University of California Press, 1999.

Boston Journal. 'Edison's Last.' May 13, 1889.

Braun, Marta, Charlie Keil, Rob King, Paul Moore, and Louis Pelletier, eds. *Beyond the Screen: Institutions, Networks and Publics of Early Cinema.* New Barnet, UK: John Libbey, 2012.

Brey, Philip. 'Technology as Extension of Human Faculties.' In *Metaphysics, Epistemology and Technology,* edited by Carl Mitcham, 59-78. London: Elsevier/JAI Press, 2000.

Burgess, Jean and Joshua Green. *YouTube: Online Video and Participatory Culture.* John Wiley & Sons, 2013.

Burns, R.W. 'The Contributions of the Bell Telephone Laboratories to the Early Development of Television.' *History of Technology* 13 (1991): 181-213.

Burns, R.W. *Television: An International History of the Formative Years.* London: Institutions of Electrical Engineers, 1998. Canales, Jimena. *A Tenth of a Second: A History.* Chicago, IL: University of Chicago Press, 2010.

Carlson, Bernard and Michael E. Gorman. 'Understanding Invention as a Cognitive Process: The Case of Thomas Edison and Early Motion Pictures, 1888-91.' *Social Studies of Science* 20, no. 3 (1990): 387-430.

Checkland, Peter. *Systems Thinking, Systems Practices.* New York: John Wiley, 1981.

Chicago Journal. 'What Edison Claims.' May 13, 1891.

Chun, Wendy. 'Introduction: Did Somebody Say New Media?' In *New Media, Old Media: A History and Theory Reader,* edited by Wendy Chun and Thomas Keenan. New York: Routledge, 2006.

Collins, Theresa, Lisa Gitelman, and Gregory Jankunis. *Thomas Edison and Modern America: a Brief History with Documents.* New York: Palgrave Macmillan, 2002.

Corwin, Sharon. 'Picturing efficiency: Precisionism, Scientific Management, and the Effacement of Labor.' *Representations* 84, no. 1 (2003): 139-165.

Croteau David and William Hoynes. *Media/ Society: Industries, Images, and Audiences.* Thousand Oaks, CA: Sage, 2014.

Darnton, Robert. 'The Symbolic Element in History.' *The Journal of Modern History* 58, no. 1 (1986): 218-234.

Darnton, Robert. 'What is the History of Books?' *Daedalus* (1982): 65-83.

Daston, Lorraine, and Peter Galison. *Objectivity*. Cambridge, MA: MIT Press, 2007.

De Laurentis, Teresa, Andreas Huyssen and Kathleen M. Woodward, ed. *The Technological Imagination: Theories and Fictions*. Madison, WI: Coda Press, 1980.

Doane, Mary Ann. *The Emergence of Cinematic Time: Modernity, Contingency, the Archive*. Cambridge, MA: Harvard University Press, 2002.

Dorson, Richard, ed. *Folklore and Folklife*. Chicago, IL: University of Chicago Press, 1972.

The Electrician, 'The Electroscope,' letter to the editor, *New York Sun,* March 30, 1877.

Elkins, James. *Visual Studies: A Skeptical Introduction*. New York: Psychology Press, 2003.

Elsaesser Thomas. 'The New Film History as Media Archaeology.' *Cinémas: Revue d'études cinématographiques Cinémas:/Journal of Film Studies* 14, no. 2-3 (2004): 75-117.

Ernst, Wolfgang. *Digital Memory and the Archive.* Edited by Jussi Parikka. Minneapolis: University of Minnesota Press, 2013.

Fickers, Andreas. 'Television.' In *The Handbook of Communication History*, edited by Peter Simonson, Janice Peck, Robert T. Craig, and John Jackson, 239-255. New York: Routledge, 2013.

Figuier, Louis. 'Le Telectroscope,' in *L'Annee Scientifique et Industrielle* (1877).

Finkelstein, David, and Alistair McCleery, ed. *The Book History Reader*. New York: Routledge, 2002.

Foucault, Michel. *The Archaeology of Knowledge*. New York: Pantheon, 1972.

Fun. 'Professor Goaheadison's Latest.' July 3, 1889.

Gaudreault, André, ed. *American Cinema, 1890-1909: Themes and Variations.* New Brunswick, NJ: Rutgers University Press, 2009.

Gaudriault, Andre. *Film and Attraction: From Kinematography to Cinema*. Chicago: University of Illinois Press, 2011.

Gernsback, Hugo. 'Television and the Telephot.' *Electrical Experimenter* (May 1918).

Gitelman, Lisa. *Always Already New: Media, History and the Data of Culture.* Cambridge, MA: MIT press, 2006.

Gitelman, Lisa. *Scripts, Grooves and Writing Machines: Representing Technology in the Edison Era.* Stanford, CA: Stanford University Press, 1999.

Gooday, Graeme. *Domesticating Electricity: Technology, uncertainty and gender, 1880–1914*. London: Pickering & Chatto, 2008.

Gooday, Graeme. *The Morals of Measurement: Accuracy, Irony, and Trust in Late Victorian Electrical Practice.* Cambridge, UK: Cambridge University Press, 2004.

Grau, Oliver, and Thomas Veigl. *Imagery in the 21st Century*. Cambridge, MA: MIT Press, 2011.

Gunning, Tom. 'An Aesthetic of Astonishment: Early Film and the (In)credulous Spectator.' *Art and Text* 34 (Spring 1989): 31-45.

Gunning, Tom. 'The Cinema of Attraction.' *Wide Angle* 3, no. 4 (1986).

Gunning, Tom. ''Primitive' Cinema: A Frame-up? Or the Trick's on Us.' *Cinema Journal* 28, no. 2 (1989): 3-12.

Hall, Stuart, ed. *Representation: Cultural Representations and Signifying Practices.* New York: Sage, 1997.

Hansen, Mark B.N. 'Media Theory.' *Theory, Culture & Society* 23, no. 2-3 (2006): 297-306.

Hansen, Mark B.N. and W.J.T. Mitchell, eds. *Critical Terms for Media Studies.* Chicago, IL: University of Chicago Press, 2010.

Haraway, Donna. *Simians, Cyborgs and Women: The Reinvention of Nature.* New York: Routledge, 1991.

Hatch, Mary Jo and Michael Owen Jones. 'Photocopylore at Work: Aesthetics, Collective Creativity and the Social Construction of Organizations.' *Studies in Cultures, Organizations and Societies* 3, no. 2 (1997): 263-287.

Hayles, N. Katherine. 'Virtual Bodies and Flickering Signifiers.' *October* 66 (1993): 69-91.

Hayles, N. Katherine. *How We Became Posthuman: Virtual Bodies in Cybernetics, Literature, and Informatics.* Chicago, IL: University of Chicago Press, 1999.

Hendricks, Gordon. *The Edison Motion Picture Myth.* Berkeley: University of California Press, 1961.

Hendricks, Gordon. *The Kinetoscope: America's First Commercially Successful Motion Picture Exhibitor.* New York: Beginnings of the American Film, 1966.

Hughes, Thomas. *Networks of Power: Electrification in Western Society, 1880-1930.* Baltimore, MD: Johns Hopkins University Press, 1993.

Hughes, Thomas. 'The Evolution of Large Technological Systems.' In *The Social Construction of Technological Systems: New Directions in the Sociology and History of Technology,* edited by Trevor Pinch, Thomas Hughes and Wiebe Bijker. Cambridge, MA: MIT Press, 1987.

Huhtamo, Erkki. 'Seeing at a Distance: Toward an Archaeology of the Small Screen.' In *Art@Science,* edited by Christa Sommerer and Laurent Mignonneau, 262-278. New York: Springer, 1998.

Huhtamo, Erkki. 'Toward a History of Peep Practice.' *A Companion to Early Cinema,* edited by André Gaudreault, Nicolas Dulac and Santiago Hidalgo, 32-51. New York: Wiley, 2012.

Jenkins, Henry. 'What Happened Before Youtube?' In *YouTube: Online Video and Participatory Culture,* edited by Jean Burgess and Joshua Green. New York: John Wiley & Sons, 2013.

Johnston, Sean. *A History of Light and Colour Measurement: science in the shadows.* Bristol, UK: Institute of Physics Publishing, 2001.

Jordan, John M. *Machine-Age Ideology: Social Engineering and American Liberalism, 1911-1939.* Chapel Hill: University of North Carolina Press, 1994.

Kern, Stephen. *The Culture of Time and Space, 1880-1918.* Cambridge, MA: Harvard University Press, 1983.

Kittler, Friedrich. *Gramaphone, Film, Typewriter.* Stanford, CA: Stanford University Press, 1999.

Kluitenberg, Eric, ed. *The Book of Imaginary Media: Excavating the Dream of the Ultimate Communication Medium.* Rotterdam, The Netherlands: NAi Publishers, 2007.

Kluitenberg, Eric. 'On the Archaeology of Imaginary Media.' In *Media Archaeology: Approaches, Applications, and Implications,* edited by Erkki Huhtamo and Jussi Parikka, 48-69. Berkeley: University of California Press, 2011.

Lastra, James. *Sound Technology and the American Cinema: Perception, Representation, Modernity.* New York: Columbia University Press, 2000.

Leary, Patrick. *The Punch Brotherhood: Table Talk and Print Culture In Mid-Victorian London.* London: British Library, 2010.

Lightman, Bernard. *Victorian Popularizers of Science: Designing Nature for New Audiences.* Chicago, IL: University of Chicago Press, 2009.

Lotz, Amanda. *The Television Will be Revolutionized.* New York: NYU Press, 2014.

Luckiesh, Matthew. 'The Human Seeing-Machine.' *Transactions of the Illuminating Engineering Society* 27 (1932): 699-722.

Maltby, Richard. 'New Cinema Histories.' In *Explorations in New Cinema History: Approaches and Case Studies,* edited by Richard Maltby, Daniel Biltereyst, and Philippe Meers, 3-40. Malden, MA: Wiley-Blackwell, 2011.

Mills, Mara. 'Deafening: Noise and the Engineering of Communication in the Telephone System.' *Grey Room* 43 (2011): 118-143.

Mindell, David A. *Between Human and Machine: Feedback, Control, and Computing Before Cybernetics.* Baltimore, MD: Johns Hopkins University Press, 2002.

Mingers, John. *Realising Systems Thinking: Knowledge and Action in Management Science.* New York: Springer, 2006.

Mitcham, Carl. *Thinking Through Technology: The Path Between Engineering and Philosophy.* Chicago, IL: University of Chicago Press, 1994.

Mitchell, Timothy. 'The World as Exhibition.' *Comparative Studies in Society and History* 31, no. 2 (1989).

Mitchell, W.J.T. *What Do Pictures Want?: The Lives and Loves of Images.* Chicago, IL: University of Chicago Press, 2005.

Monteiro, Lois A. 'The Electronic Pocket Calculator: Joke 1.' *Western Folklore* 35, no. 1 (1976): 75.

Morus, Iwan Rhys. "The Nervous System of Britain': Space, Time and The Electric Telegraph in the Victorian Age.' *The British Journal for the History of Science* 33, no. 4 (2000): 455-475.

Mumford, Lewis. *Technics and Civilization.* New York: Harcourt, 1934.

Mussel, James. *Science, Time and Space in the Late Nineteenth-Century Periodical Press*. Burlington, VT: Ashgate, 2007.

Musser, Charles *Before the Nickelodeon: Edwin S. Porter and the Edison Manufacturing Company.* Berkeley, CA: University of California Press, 1991.

Musser, Charles. *The Emergence of Cinema: The American Screen to 1907.* Berkeley, CA: University of California Press, 1994.

Natale, Simone and Gabriele Balbi, 'Media and the Imaginary in History: The Role of the Fantastic in Different Stages of Media Change.' *Media History* 20, no. 2 (2014): 203-218.

Nead, Lynda. *Myths of Sexuality: Representations of Women in Victorian Britain.* Oxford: Blackwell, 1988.

Neubauer, John, ed. *Cultural History after Foucault*. Piscataway, NJ: Transaction Publishers, 1999.

New York Sun. 'Edison's 'Ear Telescope'.' June 8, 1878.

New York Tribune. 'Untitled' (This discovery mania). May 14, 1878.

Noakes, Richard. 'Punch and Comic Journalism in Mid-Victorian Britain.' In *Science in the Nineteenth-century Periodical: Reading the Magazine of Nature*, edited by Geoffrey Cantor. Cambridge, UK: Cambridge University Press, 2004.

Noakes, Richard. 'Science in Mid-Victorian *Punch*.' *Endeavour* 26, no. 3 (2002): 92-96.

Noble, David F. *America by Design: Science, Technology, and the Rise of Corporate Capitalism.* Cambridge, UK: Oxford University Press, 1979.

Orvell, Miles. *The Real Thing: Imitation and Authenticity in American Culture, 1880-1940.* Chapel Hill: University of North Carolina Press, 1989.

Otter, Chris. *The Victorian Eye: A Political History of Light and Vision in Britain, 1800-1910.* Chicago: University of Chicago Press, 2008.

Parikka, Jussi. 'Operative Media Archaeology: Wolfgang Ernst's Materialist Media Diagrammatics.' *Theory, Culture & Society* 28, no. 5 (2011): 52-74;

Perkowitz, Sydney. *Empire of Light: A History of Discovery in Science and Art.* New York: Henry Holt, 1996.

Perskyi, Constantin. 'Television Using Electricity.' In *Congres International D. Electricite* (*International Congress on Electricity of 1900*). Paris: Gauthier –Villars, 1901.

Peters, Benjamin. 'And Lead Us Not Into Thinking the New is New: a Bibliographic Case for New Media History.' *New Media & Society* 11, no. 1-2 (2009): 13-30.

Poster, Mark. 'The Digital Subject and Cultural Theory.' In *What's the Matter with the Internet.* Minneapolis: University of Minnesota Press, 2001.

Preston, Michael. 'Xerox-lore.' *Keystone Folklore* 19, no. 1 (1974): 11-26.

Preston, Michael. 'Traditional Humor from the Fax Machine: All of a Kind.' *Western Folklore* 53, no. 2 (1994): 147-169.

Rabinbach, Anson. *The Human Motor: Energy, Fatigue, and the Origins of Modernity.* Berkeley: University of California Press, 1992.

Rasmussen, Nicolas. *Picture Control: The Electron Microscope and the Transformation of Biology in America, 1940-1960*. Stanford, CA: Stanford University Press, 1999.

Robida, Albert. *Le Vingtième Siècle* (*The Twentieth Century*). Paris: Librairie Illustree, 1882; Middletown, CT: Wesleyan University Press, 2004.

Rubinstein, Charles. 'Optics at Bell Laboratories – General Optics, television, and Vision.' *Applied Optics* 11, no. 11 (1972): 2401-2411.

Sconce, Jeffrey. *Haunted Media: Electronic Presence from Telegraphy to Television*. Durham, NC: Duke University Press, 2000.

Secord, James. *Victorian Sensation: The Extraordinary Publication, Eeception, and Secret Authorship of 'Vestiges of the Natural History of Creation.'* Chicago, University of Chicago Press, 2003.

Secord, James. *Visions of Science: Books and Readers at the Dawn of the Victorian Age*. Chicago, IL: University of Chicago Press, 2015.

Shiers, George. *Early Television: A Bibliographic Guide*. London: Taylor & Francis, 1997.

Shiers, George. 'The Rise of Mechanical Television, 1901–1930.' *SMPTE Journal* 90, no. 6 (1981): 508-521.

Sorin-George, Toma, Ana-Maria Grigore, and Paul Marinescu. 'The Emergence of Scientific Management in America.' *Manager* 19 (2014): 128-131.

Snickars, Pelle, and Patrick Vonderau, eds. *The YouTube Reader*. Stockholm: National Library of Sweden, 2009.

Spehr, Paul. *The Man Who Made Movies: W.K.L. Dickson*. New Barnet, UK: John Libbey Publishing, 2008.

Sterne, Jonathan. *The Audible Past: Cultural Origins of Sound Reproduction*. Durham, NC: Duke University Press, 2003.

Sterne, Jonathan. *MP3: The Meaning of a Format*. Durham, NC: Duke University Press, 2012.

Stiegler, Bernard. *Technics and Time: The Fault of Epimetheus*. Stanford, CA: Stanford University Press, 1998.

Strauven, Wanda, ed. *The Cinema of Attractions Reloaded*. Amsterdam: Amsterdam University Press, 2006.

Sturken, Marita, Douglas Thomas, and Sandra Ball-Rokeach. *Technological Visions: The Hopes and Fears that Shape New Technologies*. Philadelphia, PA: Temple University Press, 2004.

Telotte, J.P. *A Distant Technology: Science Fiction Film and the Machine Age*. Middletown, CT: Wesleyan University Press, 1999.

Thomas, Julia. *Pictorial Victorians: The Inscription of Values in Word and Image*. Athens: Ohio University Press, 2004.

Thompson, Kenneth. 'Introduction to the Early Sociology of Management and Organizations.' In *Scientific Management*. New York: Routledge, 2003.

Thorburn, David and Henry Jenkins, ed. *Rethinking Media Change: The Aesthetics of Transition*. Cambrige, MA: MIT Press, 2003.

Turkle, Sherry. *Alone Together: Why We Expect More from Technology and Less from Each Other.* New York: Basic Books, 2011.

Uricchio, William. 'The Future of a Medium Once Known as Television.' In *The YouTube Reader,* edited by Patrick Vonderau and Pelle Snickars, 24-39. Stockholm: National Library of Sweden, 2009.

Uricchio, William. 'Historicizing Media in Transition.' In *Rethinking Media Change: The Aesthetics of Transition,* edited by David Thorburn and Henry Jenkins, 23-38. Cambridge, MA: MIT Press, 2003.

Uricchio, William. 'Phantasia and Technè at the Fin-de-siècle.' *Intermediality: History and Theory of the Arts, Literature and Technologies* 6 (2005): 27-42.

Uricchio, William. 'Storage, Simultaneity, and the Media Technologies of Modernity.' In *Allegories of Communication: Intermedial Concerns from Cinema to the Digital,* edited by John Fullerton and Jan Olsson, 123-138. Bloomington: Indiana University Press, 2004.

Uricchio, William. 'Television, Film and the Struggle for Media Identity.' *Film History* 10, no. 2 (1998): 118-127.

Uricchio, William. 'Television's First Seventy-Five Years: The Interpretive Flexibility of a Medium in Transition.' In *The Oxford Handbook of Film and Media Studies,* edited by Robert Kolker, 286v-305. Cambridge, UK: Oxford University Press, 2008.

Uricchio, William and Roberta Pearson. 'Coming to Terms with New York City's Moving Picture Operators, 1906–1913.' *The Moving Image: The Journal of the Association of Moving Image Archivists* 2, no. 2 (2002): 73-93.

Van den Ende, Jan, Wim Ravesteijn, and Dirk De Wit. 'Shaping the Early Development of Television.' *Technology and Society Magazine, IEEE* 16, no. 4 (1997): 13-26.

Van der Vleuten, Erik. 'Large Technical Systems.' In *A Companion to the Philosophy of Technology,* edited by Jan Olsen, Stig Andur Pedersen, and Vincent F. Hendricks, 218-222. Malden, MA: John Wiley & Sons, 2012.

Van Loon, Joost. *Media Technology: Critical Perspectives*. New York: McGraw Hill, 2008.

Varnava, Andrekos. 'Punch and the British Occupation of Cyprus in 1878.' *Byzantine and Modern Greek Studies* 29, no. 2 (2005): 167-186.

Willis, Martin. *Vision, Science and Literature, 1870-1920: Ocular Horizons.* New York: Routledge, 2015.

Wilson, Catherine. *The Invisible World: Early Modern Philosophy and the Invention of the Microscope*. Princeton, NJ: Princeton University Press, 1995.

Winner, Langdon. 'Upon Opening the Black Box and Finding It Empty: Social Constructivism and the Philosophy of Technology.' *Science, Technology, & Human Values* 18, no. 3 (1993): 362-378.

Wurtzler, Steve. *Electric Sounds: Technological Change and the Rise of Corporate Mass Media.* New York: Columbia University Press, 2007.

Wythoff, Grant. 'Pocket Wireless and the Shape of Media to Come, 1899–1922.' *Grey Room* 51 (2013): 40-63.

Yuste, Antonio Perez, 'La televisión mecánica' (The Mechanical Television). In *Detrás de la Cámara. Historia de la Televisión y de sus Cincuenta Años en España* (*Behind the Scenes: History of Television and its Fifty Years in Spain*), 65-82. Madrid: Colegio Oficial de Ingenieros de Telecomunicación, 2008.

Zajonc, Arthur. *Catching the Light: The Entwined History of Light and Mind.* Cambridge, UK: Oxford University Press, 1995.

Zeise, Tina. *Geschichte und Technik des analogen Fernsehens* (*History and technology of Analog Television*). Munich: GRIN Verlag, 2006.

1. The Telephonoscope:

How a Satire of Electric Light became a Visual Telephone

Abstract
This chapter investigates the circumstances surrounding George Du Maurier's illustration of 'Edison's Telephonoscope' published in *Punch* in December 1878. The telephonoscope emerged in the context of late nineteenth-century 'discovery mania'. Media coverage on both sides of the Atlantic met Edison's inventions with a mixture of zeal and scepticism. Contemporary readers would have been acutely aware of Edison's invention of the telephonoscope (ear telescope or megaphone) and his claims to the invention of electric light that fall. I introduce the concept of 'technological folklore' to account for the rumours, hearsay, and journalist commentary that contributed to the construction of cultural representations of the telephonoscope and electric light. 'Edison's Telephonoscope' represents 'discovery mania' by negotiating between the exaggerated claims of invention and the satirical rejection of new technology for its own sake. This chapter encourages media historians to weigh the presentist perspective, which associates 'Edison's Telephonoscope' with a television or electronic screen, with the view of the contemporary reader, who would have made sense of the depiction not as a prophecy but as a speculation and critique of technology.

Keywords: *Punch*; Thomas Edison; Victorian Age; technological folklore

At first glance, George du Maurier's illustration 'Edison's Telephonoscope' from *Punch* (Figure 1), appears to picture what television would look like 100 years hence.[1] In the tradition of the best Victorian science fiction, it speculates about the look and experience of the TV screen as a kind of

1 George Du Maurier, 'Edison's Telephonoscope,' *Punch*, 9 December 1878.

Roberts, I., *Visions of Electric Media: Television in the Victorian and Machine Ages*.
Taylor & Francis Group, 2019
DOI 10.5117/9789462986596_CH01

telephone. In 'Edison's Telephonoscope', Du Maurier shows a big screen television-telephone facilitating an intimate communication between a wealthy couple in Wilton Place, London (a wealthy district home to aristocrats and politicians), and their daughter in British Ceylon (now Sri Lanka). It was said to allow family and loved ones to stay in contact despite being separated by long distances.

But what would you take away from this illustration if you had never seen a television before? In the context of the late nineteenth-century technological revolution, imagine what a contemporary Victorian reader would make of it. Alexander Graham Bell had only recently begun to demonstrate his remarkable new talking telegraph (telephone). Thomas Edison had rocked the modern world with his invention of the phonograph and the electric light. By associating the telephonoscope, a speculative apparatus, with the character of American inventor Thomas Edison, George Du Maurier's illustration satirizes the way each of these new inventions upped the ante. It signifies the absurdity of futuristic technological progress and insinuates a sceptical attitude about technological supersession.

For the past 30 years, scholars have consistently located 'Edison's Telephonoscope' within the prehistory of television and motion pictures. It has become an unexamined marker at the beginning of a trajectory that leads to the modern media landscape. But jumping to the assumption that Du Maurier forecast television neglects the many different ways the image could convey meaning. Even though *Punch*'s 'Edison's Telephonoscope' appears to resemble television to us today, the magazine's satirical style welcomes multiple interpretations. The 'Telephonoscope' can be understood as a futuristic speculation about television as well as a critical reflection on technological progress.

This chapter argues for a contextual reading of 'Edison's Telephonoscope'. From the perspective of a nineteenth-century reader of *Punch*, the picture would have carried different meaning than it does for us today, resonating with current events and references circulating in British and American culture. Considering different interpretations of 'Edison's Telephonoscope' makes it possible to relate it to scholarly literature across disciplines. An interdisciplinary approach begs the question of how to define the object under study. Depending on who you ask, the 'Telephonoscope' might be identified as a rhetorical object, a novel invention, or a satirical illustration.

The history of technology and print culture provide alternative approaches to media history. While a media archaeologist might identify 'Edison's Telephonoscope' as imaginary media, a scholar of print culture would be more likely to associate the illustration within the history of *Punch*, Victorian satire, and science fiction.[2] Treating the 'Telephonoscope' as a satirical illustration rather than a Victorian television demands that its context be recognized. Julia Thomas's analysis of representations of the crinoline in *Pictorial Victorians*, for example, demonstrates such an approach.[3]

The interdisciplinary field of literature and science also offers a way of thinking about the 'Telephonoscope' as a part of late nineteenth-century technological imagination.[4] This approach, however, encourages scholars to collapse distinctions between science and technology on the one hand and fiction on the other. A contextual reading acknowledges relationships between the 'Telephonoscope' and Edison's other inventions, and between the cultural climate of late nineteenth-century 'discovery mania' and a contemporary reader's frame of reference for *Punch*. For example, Victorian and visual-culture scholars have shown how much *Punch*'s satirical tone worked to construct meaning for its readers. Whether in a political cartoon or a joke about women's clothing, *Punch* rarely published simply whimsical cartoons. They often carried many layers of meaning: inside jokes related to current events targeted at the London reader.[5]

The concept of *technological folklore* suggests a way to bridge the differences between media, technology, and print. Media historian Lisa Gitelman defines technological folklore as 'a word-of-mouth culture of technological possibility every bit as tenacious as other folkloric traditions'.[6] While the term has been thrown around very casually in media history and history of technology scholarship, folklorists have been studying it for decades.[7] From rumours people told about the coming of the electric light in the 1870s to jokes transmitted by fax machine in the 1970s, technological folklore communicated

2 Natale and Balbi, 'Media and the Imaginary in History'.

3 Thomas, *Pictorial Victorians*.

4 Willis, *Vision, Science and Literature*; Hunt, 'Electric Leisure'.

5 Codell, 'Imperial Differences and Culture Clashes'; Thomas, *Pictorial Victorians*; Colin A. Hempstead, 'Representations of Transatlantic Telegraph,' *Engineering Science & Education Journal* 4, no. 6 (1995): 17-25; Andrekos Varnava, 'Punch and the British Occupation of Cyprus in 1878,' *Byzantine and Modern Greek Studies* 29, no. 2 (2005): 167-186; Altick, *Punch*.

6 Gitelman, *Scripts and Grooves*, 77; Darnton, *The Great Cat Massacre*, 64.

7 Marvin, *When Old Technologies were New*. See also Tom Standage, *The Victorian Internet: The remarkable story of the telegraph and the nineteenth century's online pioneers* (London: Weidenfeld & Nicolson, 1998); Jay Clayton, *Charles Dickens in Cyberspace: The Afterlife of the Nineteenth Century in Postmodern Culture* (Cambridge, UK: Oxford University Press, 2003); Lois A. Monteiro, 'The Electronic Pocket Calculator: Joke 1,' *Western Folklore* 35, no. 1 (1976): 75; Michael Preston, 'Xerox-lore,' *Keystone Folklore* 19, no. 1 (1974): 11-26; Michael Preston, 'Traditional Humor from the Fax Machine: All of a Kind,' *Western Folklore* 53, no. 2 (1994): 147-169; Mary Jo Hatch and Michael Owen Jones, 'Photocopylore at work: Aesthetics, collective creativity and the social construction of organizations,' *Studies in Cultures, Organizations and Societies* 3, no. 2 (1997): 263-287.

hopes, dreams, fears, and anxieties of socio-technical change. The concept of technological folklore supports an interdisciplinary reading of 'Edison's Telephonoscope' by fusing interests in media history, the history of technology, and print culture.[8] New media such as illustrated magazines and telegraphs were both the conduits of information and the subjects of stories. Spreading rumours, hearsay, and hype, technology provided the conduit for speculation about new invention and facilitated the late nineteenth-century culture of technological anticipation. 'Discovery mania', as it was often called, blossomed from a mixture of hyperbolic rhetoric and scepticism about socio-technical change. Variants on several key themes resurface again and again in the literature: stories about misinterpreted messages, ghosts in the wires, electrical disturbances, jokes poking fun at customers who failed to grasp the difference between the postal service and the telegraph office, long-distance romances, and love affairs between telegraph clerks.[9] Expressions of the fears and anxieties of technological change, an emphasis on the likelihood of technological breakdown, and discrepancies between expectations and technical limitations of technology are some of the most common morals of these stories.

Additionally, an historian of technology might draw connections between the 'Telephonoscope' and other Victorian inventions for seeing by electricity.[10] For example, Bernard Carlson's work on Edison's strategy of invention helps to locate the frame of reference a reader might have used when encountering late nineteenth-century technologies.[11] Carlson has shown how Edison sought to locate combinations of existing tools, particularly through the integration of electric to mechanical devices: the electric pen, for example. This strategy also spawned contrivances like the kinetoscope ('to do for the eye what the phonograph does for the ear') and the megaphone (also referred to as an 'ear telescope' or 'sound opera glasses'). Carlson argues that Edison's strategy began with practical mechanical and electric combinations, which did not always intersect with how these tools might be useful in everyday life. While Carlson introduces the notion of

8 Robert Darnton, 'The Symbolic Element in History,' *The Journal of Modern History* 58, no. 1 (1986): 218-234; Gitelman, *Scripts and Grooves*; Carlson, 'Artifacts and Frames of Meaning'; Secord, *Visions of Science*; Beegan, *The Mass Image*; Mussel, *Science, Time and Space*; Lightman, 'The Voices of Nature': Popularizing Victorian science'; Aaron Toscano, *Marconi's Wireless and the Rhetoric of a New Technology* (New York: Springer, 2012).

9 W.J. Johnston, ed. *Lightning Flashes and Electric Dashes: A Volume of Choice Telegraphic Literature, Humor, Fun, Wit & Wisdom* (New York: W. J. Johnston, 1877); W.J. Johnston, ed. *Telegraphic Tales and Telegraphic History* (New York: W. J. Johnston, 1880).

10 R.W. Burns, 'Prophecy into Practice'; R.W. Burns, *Television;* Shiers, *Early Television: A Bibliographic Guide.*

11 Carlson, 'Artifacts and Frames of Meaning'.

frames of reference to explain Edison's strategy of invention, I will use it to explain how the general public in America and Great Britain might have understood his language of invention in different ways.

This chapter dismantles assumptions about the relationship between 'Edison's Telephonoscope', Edison's telephonoscope (AKA megaphone), and other new technologies born in the late nineteenth century such as electric light, the photograph, and the telephone. Broadening the definition of invention, this chapter also considers the role of imaginary technologies like the Chrysophone and Edison's Anti-Gravity underclothing.[12] The key purpose of this chapter is to provide an historical account to support a contextual reading of Du Maurier's graphic satire. With a focus on Thomas Edison's work and the popular image of the 'Wizard of Menlo Park' he projected in the press, I look closely at the events and announcements, rhetoric and opinion that circulated throughout the year of 1878 and which led up to the publication of 'Edison's Telephonoscope' in December. With a particular emphasis on the reception of two of Edison's inventions (the megaphone and the electric light), I will examine the frame of reference for both American and British audiences.

Part one examines the megaphone in the context of the late nineteenth-century technological revolution, specifically in the climate immediately following the invention of the telephone and the promotion of the phonograph. Part two follows the announcement of Edison's invention of the electric light in October 1878. In the fallout from the news exaggerated in the press on both sides of the Atlantic, *Punch* published a spread of illustrations poking fun at Edison's character. 'Edison's Telephonoscope' was among the series of satires aimed at critiquing his overblown rhetoric, encouraging readers to think twice before tossing away their gas lamps and candlesticks. From the perspective of the Victorian reader, 'Edison's Telephonoscope' satirized the inventor's misplaced confidence. It encouraged readers to examine both the benefits and drawbacks of technological progress and supersession.

Discovery Mania: The Megaphone

1878, a year that began with demonstrations of the telephone and phonograph, inaugurated a period of hype and speculation related to the promise of new technology. Riding on the coat-tails of Alexander Graham Bell's invention of the talking telegraph (or telephone), American inventor Thomas

12 George Du Maurier, 'Recent Improvements in Science' (The Chrysophone), *Punch*, 14 December 1878; George Du Maurier, 'Edison's Anti-Gravity Underclothing, *Punch*, 9 December 1878.

Alva Edison held public demonstrations to show what his new Phonograph could do. One correspondent dubbed it 'discovery mania', characterized by the sense that technology could make the world seem like it had 'turned upside down'.[13] 'The annihilation of space and time' had been repeated so often it had long since become a cliché.[14] By year's end, it seemed like the limits to useful new invention had been reached. Ironically, the cultural forces driving technological change showed no signs of letting up.

Magazines and newspapers overwhelmed their readers with new ways of lighting public spaces, powering transportation, and communicating over great distances. The success or failure of a new invention hinged on the ability of the inventor to demonstrate its revolutionary potential. Progress seemed like an onslaught of newness for its own sake, pushing out established, tried-and-true methods attached to social customs and established norms. The technological revolution impacted every aspect of life, from kitchen appliances for the domestic sphere and fashion for the socialite, to intercontinental communications networks and power supplies that revolutionized business and industry.

The phonograph provides a gauge against which to measure the public's willingness to accept the forces of technological change. Demonstrations in early 1878 proved to the public that the recording of fugitive sounds was neither a stage illusion nor a fleeting fancy. The mixture of excitement and scepticism that surrounded Edison's new invention established expectations for Edison's later inventions, including the megaphone, the electric light, and the kinetoscope.

Those who recognized the phonograph as a tool that could make life easier initiated a discussion speculating on its many possible applications in daily life. Others saw the uses of the phonograph for surveillance and mischief. For example, New York humour magazine *Puck* pictured the phonograph as a trickster. It would clandestinely capture the furious ravings of a dignified priest in order to defame him in front of his congregation.[15] Such farces gave voice to the sceptics who doubted the value of progress for its own sake. Rather, the phonograph was a tool for mischief and deception.

Scribner's and the *New York Sun* rained down praise, running stories promoting Edison's new invention as a sign of progress. Take for example the account of the phonograph bottling the voice of a famous singer: 'The electrotype cylinders thence obtained will be put into the hand organs of the

13 *New York Tribune*, 'Untitled' (This discovery mania), 14 May 1878.
14 Kern, *The Culture of Time and Space*, 214-215; Marx, *Machine in the Garden*, 194.
15 *Puck*, 'That Awful Phonograph,' 24 April 1878.

A SUGGESTION.

How much better if, instead of hirsute Italian Organ-Grinders parading our Streets, we could have fair Female Phonographers playing our best Poets in their own original Voices!

Figure 3. George Du Maurier, 'A Suggestion (Fair Female Phonographers),' *Punch*, 20 April 1878.

streets, and we shall hear the actual voice of Christine Nilsson or Miss Cady ground out at every corner.'[16] Edison's claim that his marvellous invention

16 *New York Sun*, 'Talking of Futurity,' 24 March 1878; George Bartlett Preston, 'The Telephone and the Phonograph,' *Scribner's*, April 1878, 848-858.

would supersede the street performer must have rattled the status quo. *Punch* excelled at creating unique and memorable imagery by playing to the social anxieties of the day.[17] In the case of *Punch*'s 'Fair Female Phonographers' (Figure 3), illustrator George Du Maurier combined the familiar character of the organ grinder with the exaggerated speculation that the phonograph was poised to change every aspect of modern life. As expressed in *Punch*, satire, speculation, and scepticism converged to create the distinct tone of technological folklore.

The ability of the phonograph to make the voice present in the speaker's absence made many people uneasy. Like the power of writing technology to support the illusion of presence in bodily absence, the phonograph extended and magnified the domain of one's influence. These illustrations published on the front page of the *Daily Graphic* (Figure 4), for example, emphasized these new configurations. The picture at centre shows how the mouth interfaces with the recording mechanism. The picture at top right illustrates the fear that the phonograph will intrude on a good night's sleep. These pictures show the many possible applications of the phonograph in everyday use. While the phonograph held the potential to make work life easier, it also threatened to change pleasant aspects of daily life. Illustrations in the press correspond to the expressions of fears and anxieties of change told in the stories that circulated in technological folklore.

In the search for combinations of existing tools, Edison contrived schemes for visual and sonic adjuncts: tools to enhance hearing, mechanisms to amplify the voice, devices to capture fugitive sounds and preserve them for posterity. In this context, Edison responded to queries about the rumour that he was working on an artificial ear or apparatus for the deaf.[18]

> Some weeks ago a reporter came to see my phonograph and went back and got it all mixed up in his paper. He stated that I had got up a machine to make partially deaf people hear. The item was extensively copied, but I thought nothing more of it until after a while I found myself receiving letters from all over the country asking about it. I answered some saying it was a mistake but they kept piling in upon me until I was getting them at the rate of twenty and thirty a day. Then I began thinking about the matter and began experimenting [....]That was the first of the megaphone.[19]

17 Altick, *Punch*; Leary, *The Punch Brotherhood*; Noakes, 'Punch and Comic Journalism'; Noakes, 'Representing "a Century of Inventions"'; Noakes, 'Science in Mid-Victorian *Punch*'.
18 *New York Tribune*, 'Untitled' (This discovery mania), 14 May 1878.
19 *Illustrated Science News*, 'Edison's Marvelous Inventions,' 1 July 1878, 103.

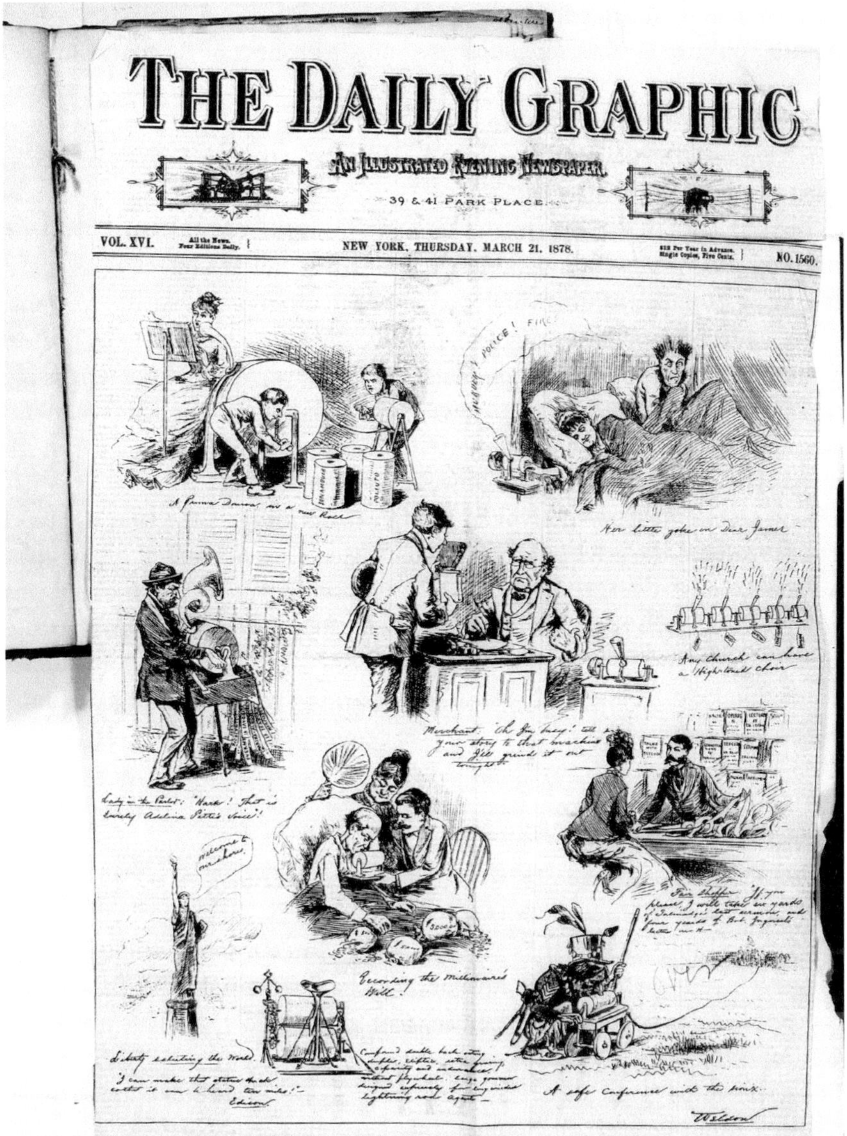

Figure 4. 'Awful Possibilities of the New Speaking Phonograph,' *Daily Graphic*, 21 March 1878.

Some called it an ear trumpet or a hearing aid. Edison initially called it the *telephonoscope*, a tool for magnifying sounds over a distance. When Menlo Park technician James Redpath jokingly referred to it as a *telescopophone*, Edison sought out different indicators of its practical value. He began calling it a megaphone, an ear telescope, or 'sound opera glasses'. It made its way into popular culture as the *telephonoscope*, more a product of 'discovery mania' than of the mind of Edison. Initially, as indicated by Redpath's derision, referring to the megaphone as a *telescopophone* targeted the perceived *uselessness* of the device. It was cumbersome: too big to be of any practical use.

As indicated in Edison's initial sketches from March and April 1878, the telephonoscope was intended to merge the functions of a megaphone and a telescope (Figure 2). Edison envisioned two distinct practical applications. First, 'sound opera glasses' would amplify the performances in a theatre. Second, it would be an advantageous tool for surveyors, making it possible to communicate verbally as well as visually across the land.[20] The two initial purposes assigned to the megaphone attracted attention from readers both optimistic and sceptical of the power of new technology.

The *telephonoscope* conveyed the overwhelming sense that it would not actually make anybody's life much easier. This perceived *uselessness* helped to construct the telephonoscope as a symbol of everything that was going wrong with the technological revolution. A demonstration for the *Daily Graphic* served to confirm many of these fears. Menlo Park assistants Batchelor, Painter, and Redpath set up a pair of megaphones at a distance of 600 feet:

> One of them said, 'Do you hear me now?'
> 'Yes!' shouted they in the porch, with a laughing accompaniment, for the voice was distinctly audible to all with the naked ear.[21]

Demonstrations such as these attracted unwanted attention, showing off the new invention like a silly curio. Edison had promised a device to make it possible to hear farther and with greater amplification. This unwieldy contraption hardly met the expectations set up by Edison's grand claims. Redpath jokingly called it a *telescopophone*. The *New York Sun* chimed in, 'as a voice cannot be seen, the name is incongruous and absurd'.[22] An *English Mechanic* reader wrote in to the paper: 'This instrument is so unwieldy and of so little value that it is not worth making.'[23] But Edison, in an attempt to

20 *New York Sun*, 'Edison's "Ear Telescope",' 8 June 1878.

21 *The Daily Graphic*, 'Ears for the Deaf,' 5 June 1878.

22 *New York Sun*, 'Edison Should Make Haste to Abandon the Terrific Name' (telescopophone), 11 June 1878.

23 W. J. Lancaster, 'Edison's Megaphone,' letter to the editor, *English Mechanic*, 24 January 1879, 501.

retain a sense of dignity, forced on it the designation of 'Megaphone: the great sounder'.[24]

While for the deaf it could potentially supplement the faculty of hearing, for the vast majority of the population, the telephonoscope supported the sceptic's stance that new technologies did nothing but enhance the power of those who already possessed it. A tool for hearing at a distance would privilege only those who could afford something so frivolous. News stories highlighted potential uses of surveillance, showing how the cumbersome machine would be best suited as a party trick for spreading gossip. Echoing the fears that the phonograph would bottle the voice, similar rumours circulated about the telephonoscope. For example, the *Indianapolis Herald* warned that Edison's invention of the telephonoscope should be taken with 'considerable license':

> Persons who were seriously alarmed lest the phonograph was destined to invade all privacy and furnish the very walls with ears, can rest easy for a little while, until it grows to be a more dreadful machine than it is.[25]

Already in June 1878, journalists had caught on to the breach between Edison's claims and his actual accomplishments. Shortly after the idea of the telephonoscope coalesced, it became clear that it would be used for purposes far removed from the polite circles of opera and the practicalities of land surveying. *Scribner's* illustrator James E. Kelly pictured the megaphone as a surveillance device, used for collecting gossip (Figure 5). *Scientific American*'s illustration of the megaphone, featured on the front page of an August issue (Figure 6), also hints at the likelihood that the device would be used for overhearing the conversations of neighbours. To the back porch setting chosen by Kelly, *Scientific American* added a distinctly pastoral feeling to its depiction, suggesting that the megaphone would have appealed to a particularly middle-class, country audience. Both illustrations depict a non-urban setting, with the device applied to a distinctly leisure purpose. Accompanying the exaggeratedly optimistic tone in which the article was written, the combination of word and image presented in *Scientific American*'s coverage of Edison's megaphone seems to have contributed to the increasingly perceptible divide between those who promoted progress for its own sake and those who harboured a sceptical attitude towards the revolutionary potential of new technology.

24 *New York Sun*, 'Edison's "Ear Telescope",' 8 June 1878.

25 *Indianapolis Herald*, 'The Amusement Record,' 15 June 1878.

GOSSIP, BY THE MEGAPHONE.

Figure 5. James E. Kelly, 'Gossip, by the Megaphone,' *Scribner's*, November 1878.

The *Daily Graphic* also published a graphic depiction of Edison's megaphone (Figure 7), which serves to illustrate the criticism lodged against its cumbersome and impractical size. What Edison had described as a mobile telescope-like tool for hearing at a distance turned out to be a massive and unusually complicated apparatus. Caught up in the relentless push towards innovation, the problem became less about supplementing the faculty of

EDISON'S MEGAPHONE.

Figure 6. 'Edison's Megaphone,' *Scientific American*, August 1878.

hearing than about annihilating distance. A *Tribune* journalist admitted that it would be 'audacious' of an inventor to suggest he could improve on these human abilities, while simultaneously acknowledging that the power is within our grasp.[26] Enthusiasm for Edison's new invention, the megaphone, persisted throughout the year only to die out with news of something new altogether. An even greater, more monumental invention was sure to revolutionize the world. Once October rolled around, people seemed to have forgotten about the telephonoscope because they were blinded by the electric light.

26 *New York Tribune*, 'Untitled' (This discovery mania), 14 May 1878.

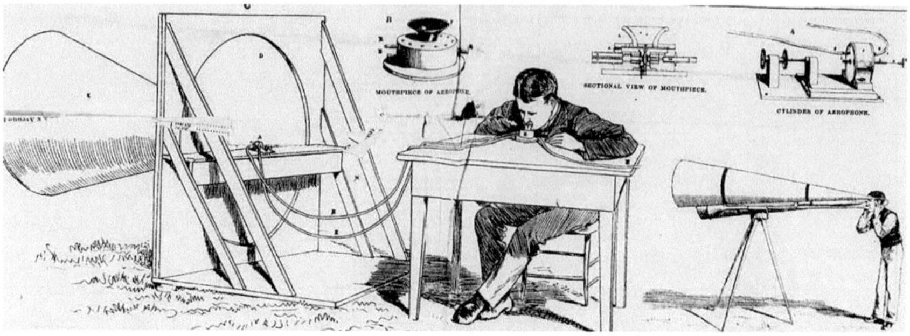

Figure 7. 'Edison's Latest Inventions; More About Edison's Wonders,' *Daily Graphic*, 19 July 1878.

Electric-Light Mania

In mid September 1878, the *New York Sun* published a series of interviews in the form of human interest stories about Edison's revolutionary experiment in electric power.[27] In it, he spoke of his intention to harness the power of Niagara. He promised to bring electric light to New York, by means of a system for subdividing and delivering electric power to domestic, public, and industrial life. These interviews invigorated public speculation and enthusiasm of all things Edison. In the U.S., the journals publicized the news with such enthusiasm that nobody really noticed when the gas stocks declined. Speculation ran rampant that the new thing of the future would be electric light, and gas would be a thing of the past. To borrow a term from historian Richard Altick, one could refer to this episode as the electric light 'mania' of 1878, in which Edison played a starring role.[28]

The *Sun* was the first to publicize Edison's interest in developing the technology to subdivide electrical power.[29] Other New York papers were quick to adopt the *Sun*'s enthusiasm for the story, which gained momentum throughout the month of September. The London papers were slightly behind the curve. It took nearly a month for the news to reach Britain. It

27 *New York Sun*, 'Invention's Big Triumph,' 10 September 1878; *New York Sun*, 'Edison's Newest Marvel,' 16 September 1878; *New York Sun*, 'Power Flashed by Wire,' 17 September 1878.

28 Altick, *Punch*, 450-492. See also Charles Bazerman, *The Languages of Edison's Light* (New York: Springer, 2002); Graeme Gooday, *Domesticating Electricity: Technology, Uncertainty and Gender, 1880–1914* (London: Pickering & Chatto, 2008); Chris Otter, *The Victorian eye: A Political History of Light and Vision in Britain, 1800-1910* (Chicago, IL: University of Chicago Press, 2008); Thomas Parke Hughes, *Networks of Power: Electrification in Western Society, 1880-1930* (Baltimore, MD: John Hopkins University Press, 1993).

29 *New York Sun*, 'Invention's Big Triumph,' 10 September 1878.

caused immediate unrest, opening up a rift between those who believed wholeheartedly in the promise of technological change and those sceptical of consequences. As a Manchester paper put it: 'It may perhaps be of value to add that the reports circulating in some quarters to the effect that "the whole thing is a hoax" must have emanated from interested parties, as there is no doubt whatever that Mr. Edison has made the "alleged" discovery in question.'[30] The British papers reflected a growing distrust of Edison's rhetoric, calling for a public demonstration to authenticate the promise. Seeing is believing, they said, unwilling to take the American journalists at their word.

Londoners identified several benefits of electricity over gas, which served as the nucleus of speculation and scepticism driving electric-light mania throughout the fall of 1878. First, electric light was thought to be cheaper and safer than gas. Second, it would have an effect of beautification, since it shines a pure light unlike the orange-tinted gas light. The discourse of technological change ignited politically acute debates related to hygiene and safety in the workplace. In its extreme, it became a matter of liberty, freeing consumers from the shackles placed on them by a corrupt gas industry. Perceptions of corrupt meter-reading practices, complaints about exorbitant and inflated prices, and fears of explosion encouraged the view that electricity would provide a cleaner, odourless, colourless, and more easily controllable alternative.

The second major aspect of the electric-light mania concerned its aesthetics. Preoccupations with the look, colour, and brightness of the coming light were reflected in the columns of the *Times* (London). A series of letters to the editor of the *Times* argues back and forth as to the way the look of the electric light would change the way people look, particularly women and particularly in public. One correspondent proclaimed that he had seen the electric light in Belgium.[31] It turned railway passengers into walking corpses, lit under a ghastly blue. The suggestion that electric light would be used in art galleries also caused controversy. As it stood, the lighting situation in London galleries was uneven. Curators depended on the location of skylights and windows mixed with gaslight that gave the impression of uneven colour and brightness. In contrast, the blue shade of the electric light would provide a superior alternative. Some agreed that the electric light would provide a superior, consistent illumination. Others feared that the blue shade of the electric light would change the look of the art. The

30 *Manchester City News*, 'The Electric Light and Mr. Edison's Discovery,' 19 October 1878.

31 M. O., letter to the editor, *Times* (London), 8 October 1878.

Figure 8. 'Brightness and Beauty,' *Punch*, 13 November 1878.

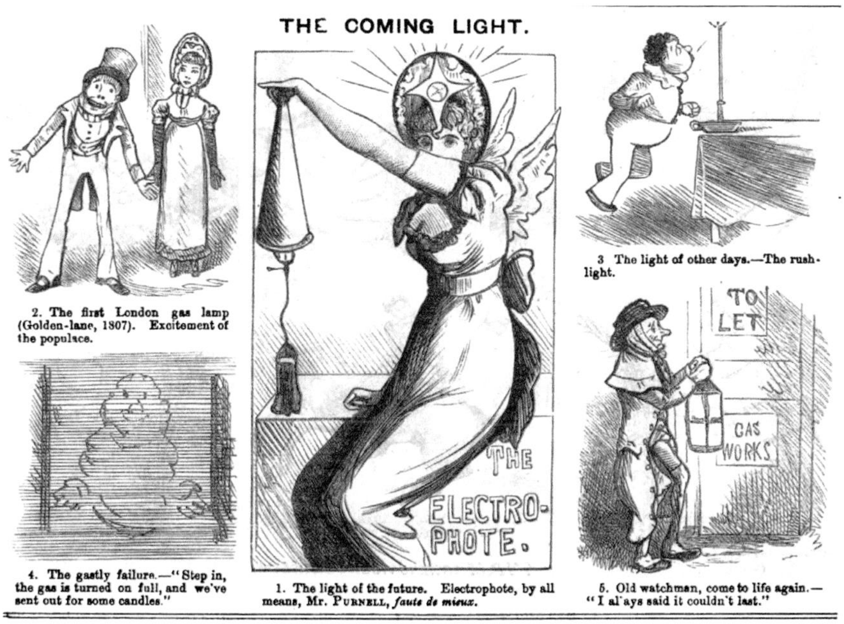

Figure 9. 'The Coming Light,' *Judy*, 24 December 1878.

aesthetics of the electric light became a matter of contrasting the shock of the blue shade against the subtler, more pleasant glow of the orange-shaded gas light, reminiscent of sunset.[32] One correspondent likened the quality of gaslight to an evening glow, noting now the female complexion is so much more pleasing in these settings. The intrusion of electric light would not only change that aesthetic, it would tarnish a woman's beauty.

The parodies and satires of the electric-light mania that appeared on the pages of *Punch* correspond with these letters in the *Times*. As Altick put it in his history of *Punch*, '[i]n a way, indeed, though nobody would have dreamed of putting it thus, *Punch* served as a weekly illustrated comic supplement of the *Times*, reflecting as in a distorting mirror a selection of the week's news and jauntily editorializing on its significance'.[33] Every week, the *Punch* contributors gathered for dinner to discuss the main theme for their next issue. As Patrick Leary describes in his history of the writing table, the humorists discussed current events and key issues over food and drink, agreeing upon a central theme around which to form each issue.[34] As a result,

32 L. W., letter to the editor, *Times* (London), 9 October 1878.
33 Altick, *Punch*, xix.
34 Leary.

their witticism was highly topical and relevant to the British audience to whom they spoke. They poked fun at public figures using caricatures and inside jokes. In some cases, the clues to deciphering their shorthand were restricted to a metropolitan audience. Altick has shown how their parodies often relied on a common language and frame of reference that would have been inaccessible to foreigners.[35] While the mirror-and-reflection metaphor might work appropriately for the condensed episode of the electric-light mania, it fails to support accurately the gravity of the events that followed. Clinging to the assumption that *Punch* reflected the cultural climate turns out to be an overly simplistic understanding. Instead, it would be beneficial to consider the representations in *Punch* as the stuff of technological folklore, which arise from and also contribute to the broader cultural perceptions of socio-technical change.

Mr. Punch was a constant critic of mercurial fads. That tone came out with full force when satirizing the vacuous claims made about the way electric light made otherwise pretty women look hideous. Punch watches surreptitiously out the window as Vanessa heads to the post office with letter in hand. Her letter, which accompanies the illustration, inquires about the effect of the electric light on her complexion (Figure 8). Stop worrying about the gas shares, she writes, and start worrying about how it will affect appearance. Similarly, a column entitled 'In the Light of the Future, Or, How we shall have to Talk', presents a dialogue making fun of the neologisms like the 'actinic halo', new accoutrements like 'head-protector' and 'pebble spectacles' one would carry to shield the eyes from the bright rays, and the speculation over how the different lights would alter one's perception of colours.[36] *Punch* articulated the dimensions of socio-technical change that were becoming recognizable in the most mundane of places, distorting the anxieties and speculations with the humorist's expertise. While the 'Telephonoscope', printed on the page opposite from 'How we shall have to Talk', makes it seem like those preoccupations were primarily visual, in the form of a screen and a darkened room, the problems rippled out into many aspects of everyday life.

Judy and *Fun* also chimed in on the electric-light controversy, noting how the debate was divided on lines of self-interest. *Judy* unveiled 'The Electrophote' (Figure 9), hailed as 'the light of the future', alongside those it would supersede: gas lamps, smoky interiors, 'the rush-light', and the

35 Altick, 92.

36 *Punch*, 'In the Light of the Future, Or, How We Shall Have to Talk,' 9 December 1878.

old watchman.[37] In its 20 November issue, *Fun* published 'The Scientific Age', a dialogue set in 'A House of the future fitted with the latest scientific appliances of the period, including telephones, phonographs, microphones, tasimeters, electro-dynamic lights, &c., &c., &c.'[38] The drama touches upon the expectations for the electrified household. By highlighting the many new technologies recently made available, alongside other speculative devices still on the horizon, the dialogue captures the atmosphere of hype and anticipation. *Fun*'s 'The Scientific Age' prefigures 'Edison's Telephonoscope' that would appear in *Punch* just two weeks later.

Punch's 9 December issue revolved around the theme of progress, innovation, and technological change particularly as it had to do with electrical power and light. *Punch*'s centrefold served to condense the issue's main idea into a single, pithy spread. The 9 December issue's centrefold, entitled 'Prometheus Unbound' (Figure 10), depicts the classic battle of man vs. God and the consequences of electric power in the hands of human inventors and industrialists. This issue served as a kind of memorial, looking back on the events that had occurred throughout the year, both fictional and real. It adopted a tone as exaggerated as that of the American literary journalists who promoted technological progress, though warped in what Altick described as a distorted mirror, in order to show the irony and hilarity of it all. *Punch* painted the exaggerated difference between the progressive rhetoric and the sceptical response in relief as only the best satire can.

Edison, the progenitor of electric-light mania, in the spotlight, became the lightning rod for criticism of technological change. Reading 'Edison's Telephonoscope' within the context of both the 9 December issue and the events that led up to its publication provides the essential frame of reference for understanding how a contemporary audience would have made sense of it. First, consider the connections a contemporary reader might have made between Edison's public character in the context of the electric-light mania and the discovery mania from earlier that summer. Edison's megaphone had been a target for criticism related to the negative consequences of technology. This critical view clashed with the exaggerated way in which the penny press was hyping up everything Edison. Subsequently, the confusion around the megaphone's name and use, entwined as it was with the many half-built and half-cocked ideas littering the Menlo Park lab, propelled the telephonoscope as an object of technological folklore.

37 *Judy, or the London Seriocomic Journal,* 'The Coming Light (Electrophote),' 24 December 1878.

38 *Fun*, 'The Scientific Age,' 20 November 1878. See also *Fun*, 'Electric Light Memo's,' 13 November 1878.

Figure 10. 'Prometheus Unbound: Science in Olympus,' *Punch*, 9 December 1878.

Du Maurier and his fellow editors drew on that technological folklore when authoring their 9 December issue. The similarities between Du Maurier's 'Telephonoscope' and *Scientific American*'s depiction of the megaphone suggests that *Punch* drew from Edison's telephonoscope in more than just the name (Figures 1, 6). Du Maurier's 'Telephonoscope' recasts *Scientific America*'s 'Megaphone' in several ways. Both depict a patriarch in the process of using the telephonoscope. Despite American and British differences, they portray what could be described as a privileged upper-class setting, particularly when noting the ways in which they both focus on depicting a social interaction. Appealing to the middle-class sensibilities of its readership, *Punch* had a reputation for poking fun at snobbish characters and characteristics, an aspect of the magazine's tone that would have been clear to a contemporary reader.[39]der, *Punch* cast the telephonoscope-user in such a role, using the technological device as a way to emphasize the social divide between the haves and the have-nots. Comparing the way *Scientific American*'s 'Megaphone' and Du Maurier's 'Telephonoscope' both illustrate the pastoral backdrop also deserves scrutiny. The American pastoral horizon depicted in *Scientific American* transforms in *Punch* into an imperialist gaze from London

39 Altick, *Punch*, 496, 566; Noakes, 'Science in Mid-Victorian *Punch*,' 92-96.

to the antipodes facilitated by technological means.[40] *Punch*'s satire implicitly asks the reader to acknowledge the social divide constructed by new technology.

Flipping through the pages of *Punch*'s 9 December issue, hurriedly printed at the end of 1878, two other illustrations corroborate the satire on technology presented in the 'Telephonosope'.[41] The issue as a whole demonstrates a sense of ambivalence towards invention and progress. Exaggerating the newness against the futuristic and frivolous served to highlight the devices' impracticality

Along with the 'Telephonoscope', 'Prometheus Unbound', and 'How we shall have to Talk', 'The Museum of Modern Antiques' and 'Edison's Anti-Gravitation Underclothing' (Figures 11, 12) delicately balance satire, scepticism, and speculation. 'Edison's Anti-gravity underclothing', a series of three illustrated panels pictures men, women, and children floating around in the air in a variety of settings: the art gallery, the park, and the nursery room. While it could be read as an outright desire for such an innovation, its whimsical frivolousness also ridicules upper-class snobbery.[42] *Punch* gave voice to the fears and anxieties of the middle class, poignantly articulating the potential havoc that promises of innovation can wreak on the existing stable economy: 'It is fast becoming obvious that unless something is done to suppress Mr. Edison, the American inventor, all the existing conditions of life will be revolutionized and *the world generally turned upside down.*'[43] As the middle class grew weary of the exaggerated rhetoric, it became more and more clear that the grandiose claims made about new invention did not always align with the improvements they promised for everyday life. The electric light held the promise of a better life. In the midst of the controversy, Londoners might have begun to wonder whether it was still worth it.

Reflecting concerns that new innovations were forcing tried-and-true techniques, practices, and occupations into obsolescence, Mr. Punch visits the 'Museum of Modern Antiques', littered with relatively new inventions such as the penny farthing, the torpedo, and the gas-powered street lamp. *Punch* played on both sides of the inexorable drive to invent. The onslaught

40 Marx, *Machine in the Garden*; E. Ann Kaplan, *Looking for the other: Feminism, film, and the imperial gaze*. New York: Routledge, 1997.

41 Marion Spielmann, *The History of Punch* (London: Cassell Publishing, 1895), 87; George Somes Layard, *The Life and Letters of Charles Samuel Keene* (London: Sampson Low, Marston and Co., 1892), 283-285; Edward Linley Sambourne, Diary (transcript) (London: The Royal Borough of Kensington and Chelsea, 1878).

42 Altick, 494.

43 *Weekly Dispatch*, 'The Electric Light,' 13 October 1878.

Figure 11. 'Museum of Modern Antiques,' *Punch*, 9 December 1878.

of newness led to the unveiling of more and more contrived mechanisms. In this graphic satire, Mr. Punch's electromagnetic hat causes his hair to stand on end, while an old sewing machine gathers dust on a nearby shelf and a gas lamppost in the background is labelled as a 'rare specimen'. The museum displays the effects of forced obsolescence. Household items like sporting goods (tennis racket and punching bag) signify the wealth of those who can afford leisure activities. In contrast, everyday household items including candles and candle snuffers, fireplace pokers and toaster lay discarded in the glass case. In addition, the museum displays a host of machines that were still relatively new in 1878, such as the Henry repeating rifle, sewing machine, and penny farthing.

Punch drew connections between technology and class, fashion fad and aesthetics that had yet to be articulated elsewhere. It acknowledged that the

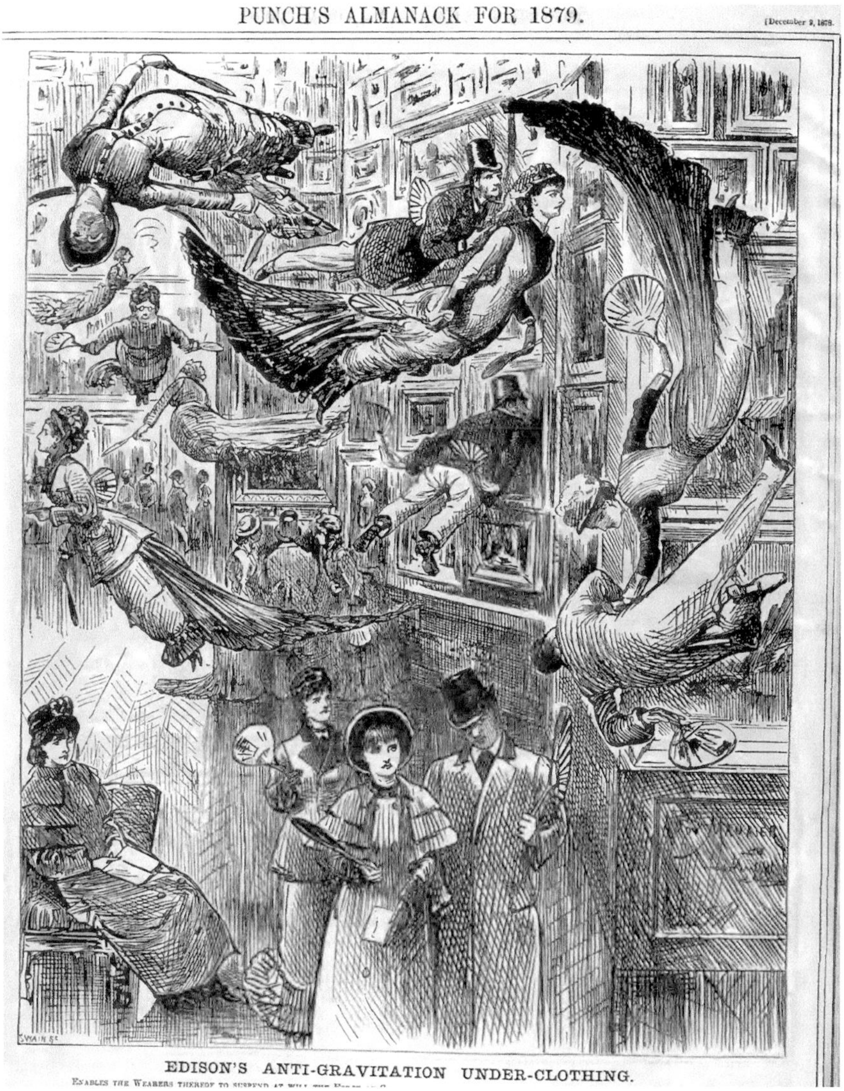

Figure 12. 'Edison's Anti-Gravitation Underclothing,' *Punch,* 9 December 1878.

British middle class might have liked to be able to have a telephonoscope in their very own sitting room. At the same time, *Punch* turned that expectation upside down. 'Edison's Telephonoscope', 'Anti-gravity Underclothing', and the 'Museum of Modern Invention' parody themes of technology and social change acutely relevant to the middle-class British readership in the fall of 1878.

These British comics chronicle a topsy-turvy year in technological change. The illustrations and articles articulated the discrepancies between the big dreams embedded in futuristic technology and the utterly improbable, which had become fused so imperceptibly in both verbal and visual discourse. *Punch* drew on cultural stereotypes like the old school master, the organ grinder, and paterfamilias to maintain a balanced editorial voice. *Punch*'s style of satire and parody enables a hermeneutic flexibility that relies heavily on the reader's frame of reference.[44]

Conclusion

Understanding 'Edison's Telephonoscope' as a satire that circulates within a cultural context displaces the assumptions that Du Maurier's picture conveys a concrete artistic intention or allows for a single reader interpretation. 'Edison's Telephonoscope' appears at first to reflect upon and satirize the electric-light mania. Read within its *circuit of culture*, the illustration assumes that the reader will recognize the reference to Edison's megaphone in its title, which references the contemporary conflict between the exaggerated promotional rhetoric and the sceptical critical backlash.

'Edison's telephonoscope' provides one of the earliest and certainly the most reprinted depictions of 'seeing by electricity'. Talk of a machine that would make it possible to see at a distance, like Bell's telephone had made it possible to convey the voice, began appearing in the spring of 1878.[45] Du Maurier visualized what a visual telephone could look like, giving concrete form to the burgeoning culture of 'seeing by electricity' and establishing its most enduring imagery.

While, for a British audience, 'Edison's Telephonoscope' parodied the electric-light mania, the story continues when the picture circulated in American culture. The image resurfaced in 1880 and again in 1891 within the context of speculation about the invention of a visual telephone and Edison's announcement of the invention of the Kinetograph, the first motion-picture camera.[46] Certainly, having a picture of what a motion-picture screen *could* look like served to unite opinion and foster expectations. Writers referenced

44 Carlson, 'Artifacts and Frames of Meaning'; Altick, 92.

45 'The Electroscope,' *New York Sun*, 30 March 1877; Louis Figuier, 'le Telectroscope,' *L'Annee Scientifique et Industrielle* 21, no. 6 (1878): 80-81; W. Donisthorpe, 'Talking Photographs,' letter to the editor, *Nature* 17, no. 430 (1878): 242; J. F. W. letter to the editor, *Nature* 18, no. 450 (1878): 169.

46 W. E. Ayrton and John Perry, 'Seeing by Electricity,' letter to the editor, *Nature* 21, no. 546 (April 1880): 589; *Illustrated American*, 'Punch and the Kinetoscope,' 20 June 1891, 224.

the *Punch* image again and again as a prophecy and as a promise, both for the coming of television and of cinema.[47] Understood as an articulation of the possibility of moving-image technology on the one hand and electrical power on the other, *Punch*'s mode of satire stresses the importance of balancing the historical and cultural contexts.

The tendency to read the 'Telephonoscope' as a Victorian form of television arises from the assumption that the picture can be read based on its appearance alone. Readjusting our frame of reference to that of a contemporary reader reveals how the satire combines current topics that occupied the minds of *Punch*'s Victorian readers: discovery mania, Edison's Telephonoscope (megaphone), and electric-light mania. Understanding *Punch*'s 'Telephonoscope' as both a speculation about moving-image technology *and* a critique of Victorian invention requires the reader to recognize the conflict between enthusiasm and scepticism. Far from a static document, it grew from and contributed to a rich technological folklore concerning progress, change, and the consequences of innovation.

Bibliography

Altick, Richard. *Punch: The Lively Youth of a British Institution, 1841-1851*. Columbus: Ohio State University Press, 1997.

Ayrton, W.E. and John Perry. 'Seeing by Electricity,' letter to the editor, *Nature* 21, no. 546 (April 1880): 589.

Bazerman, Charles. *The Languages of Edison's Light*. Cambridge, MA: MIT press, 1999.

Beegan, Gerry. *The Mass Image: A Social History of Photomechanical Reproduction in Victorian London*. London: Palgrave Macmillan UK, 2008.

Burns, R.W. 'Prophecy into Practice: The Early Rise of Videotelephony.' *Engineering Science & Education Journal* 4, no. 6 (1995): 33-40.

Burns, R.W. *Television: An International History of the Formative Years*. London: Institutions of Electrical Engineers, 1998.

Carlson, Bernard. 'Artifacts and Frames of Meaning: The Cultural Construction of Motion Pictures.' In *Shaping Technology/Building Society: Studies in Sociotechnical Change*, edited by Wiebe E. Bijker and John Law, 175-198. Cambridge, MA: MIT press, 1992.

47 Penn Steele, 'Anticipations: How some of Mr. Wells' Speculative Predecessors have Fared,' *The Era Magazine*, January-June 1902, 460-468.

Clayton, Jay. *Charles Dickens in Cyberspace: The Afterlife of the Nineteenth Century in Postmodern Culture.* Cambridge, UK: Oxford University Press, 2003.

Codell, Julie. 'Imperial Differences and Culture Clashes in Victorian Periodicals' Visuals: The Case of *Punch.*' *Victorian Periodicals Review* 39, no. 4 (2006): 410-428.

The Daily Graphic. 'Ears for the Deaf.' June 5, 1878.

Darnton, Robert. *The Great Cat Massacre and Other Episodes in French Cultural History.* New York: Basic Books, 1984.

Darnton, Robert. 'The Symbolic Element in History.' *The Journal of Modern History* 58, no. 1 (1986): 218-234.

Donisthorpe, W. 'Talking Photographs.' letter to the editor. *Nature* 17, no. 430 (1878): 242.

Figuier, Louis. 'Le Telectroscope,' in *L'Annee Scientifique et Industrielle* (1877).

Fun, 'The Scientific Age,' Nov 20, 1878.

Fun. 'Electric Light Memo's.' November 13, 1878.

Gitelman, Lisa. *Scripts, Grooves and Writing Machines: Representing Technology in the Edison Era.* Stanford, CA: Stanford University Press, 1999.

Gooday, Graeme. *Domesticating Electricity: Technology, uncertainty and gender, 1880–1914.* London: Pickering & Chatto, 2008.

Hatch, Mary Jo and Michael Owen Jones. 'Photocopylore at Work: Aesthetics, Collective Creativity and the Social Construction of Organizations.' *Studies in Cultures, Organizations and Societies* 3, no. 2 (1997): 263-287.

Hughes, Thomas. *Networks of Power: Electrification in Western Society, 1880-1930.* Baltimore, MD: Johns Hopkins University Press, 1993.

Illustrated American. 'Punch and the Kinetoscope.' June 20, 1891, 224.

Illustrated Science News. 'Edison's Marvelous Inventions.' July 1, 1878, 103.

Indianapolis Herald, 'The Amusement Record,' June 15, 1878.

J.F.W. letter to the editor, *Nature* 18, no. 450 (June 13, 1878): 169.

Johnston, W.J., ed. *Lightning Flashes and Electric Dashes: A Volume of Choice Telegraphic Literature, Humor, Fun, Wit & Wisdom.* New York: WJ Johnston, 1877.

Johnston, W.J., ed. *Telegraphic Tales and Telegraphic History.* New York: WJ Johnston, 1880.

Judy, or the London Seriocomic Journal. 'The Coming Light (Electrophote).' December 24, 1878.

Kaplan, E. Ann. *Looking for the Other: Feminism, Film, and the Imperial Gaze.* New York: Routledge, 1997.

Kern, Stephen. *The Culture of Time and Space, 1880-1918.* Cambridge, MA: Harvard University Press, 1983.

L.W., letter to the editor, *Times* (London), October 9, 1878.

Lancaster, W.J. 'Edison's Megaphone,' letter to the editor, *English Mechanic,* January 24, 1879, 501.

Layard, George Somes. *The Life and Letters of Charles Samuel Keene.* London: Sampson Low, Marston and Co., 1892.

Leary, Patrick. *The Punch Brotherhood: Table Talk and Print Culture In Mid-Victorian London.* London: British Library, 2010.

Lightman, Bernard. "The Voices of Nature': Popularizing Victorian science.' In *Victorian Science in Context,* edited by Bernard Lightman, 187-211. Chicago, IL: University of Chicago Press, 1997.

M.O., letter to the editor, *Times* (London), October 8, 1878.

Manchester City News. 'The Electric Light and Mr. Edison's Discovery.' October 19, 1878.

Marvin, Carolyn. *When Old Technologies Were New.* Cambridge, UK: Oxford University Press, 1988.

Marx, Leo. *The Machine in the Garden: Technology and the Pastoral Ideal in America.* Cambridge, UK: Oxford University Press, 1964.

Monteiro, Lois A. 'The Electronic Pocket Calculator: Joke 1.' *Western Folklore* 35, no. 1 (1976): 75.

Mussel, James. *Science, Time and Space in the Late Nineteenth-Century Periodical Press.* Burlington, VT: Ashgate, 2007.

Natale, Simone and Gabriele Balbi, 'Media and the Imaginary in History: The Role of the Fantastic in Different Stages of Media Change.' *Media History* 20, no. 2 (2014): 203-218.

New York Sun. 'Edison's 'Ear Telescope'.' June 8, 1878.

New York Sun. 'Edison's Newest Marvel.' September 16, 1878.

New York Sun. 'Edison Should Make Haste to Abandon the Terrific Name' (telescopophone). June 11, 1878.

New York Sun. 'The Electroscope.' March 30, 1877.

New York Sun. 'Invention's Big Triumph.' September 10, 1878.

New York Sun. 'Talking of Futurity.' March 24, 1878.

New York Sun. 'Power Flashed by Wire.' September 17, 1878.

New York Tribune. 'Untitled' (This discovery mania). May 14, 1878.

Noakes, Richard. 'Punch and Comic Journalism in Mid-Victorian Britain.' In *Science in the Nineteenth-century Periodical: Reading the Magazine of Nature,* edited by Geoffrey Cantor. Cambridge, UK: Cambridge University Press, 2004.

Noakes, Richard. 'Representing 'a Century of Inventions': Nineteenth-century Technology and Victorian *Punch.*' In *Culture and Science in the Nineteenth-Century Media,* edited by L. Henson, G. Cantor, G. Dawson, et al., 151-163. Farnham, UK: Ashgate, 2004.

Noakes, Richard. 'Science in Mid-Victorian *Punch.*' *Endeavour* 26, no. 3 (2002): 92-96.

Otter, Chris. *The Victorian Eye: A Political History of Light and Vision in Britain, 1800-1910*. Chicago: University of Chicago Press, 2008.

Preston, Michael. 'Xerox-lore.' *Keystone Folklore* 19, no. 1 (1974): 11-26.

Preston, Michael. 'Traditional Humor from the Fax Machine: All of a Kind.' *Western Folklore* 53, no. 2 (1994): 147-169.

Preston, George Bartlett. 'The Telephone and the Phonograph.' *Scribner's,* April 1878, 848-858.

Puck. 'That Awful Phonograph.' April 24, 1878.

Punch. 'Edison's Anti-Gravity Underclothing,' December 9, 1878.

Punch. 'Edison's Telephonoscope.' December 9, 1878.

Punch, 'In the Light of the Future, Or, How We Shall Have to Talk,' December 9, 1878.

Punch. 'Recent Improvements in Science' (The Chrysophone). December 14, 1878.

Sambourne, Edward Linley. Diary (transcript). London: The Royal Borough of Kensington and Chelsea, 1878.

Secord, James. *Visions of Science: Books and Readers at the Dawn of the Victorian Age*. Chicago, IL: University of Chicago Press, 2015.

Shiers, George. *Early Television: A Bibliographic Guide*. London: Taylor & Francis, 1997.

Spielmann, Marion. *The History of Punch*. London: Cassell Publishing, 1895.

Standage, Tom. *The Victorian Internet: The Remarkable Story of the Telegraph and the Nineteenth Century's Online Pioneers*. London: Weidenfeld & Nicolson, 1998.

Thomas, Julia. *Pictorial Victorians: The Inscription of Values in Word and Image*. Athens: Ohio University Press, 2004.

Toscano, Aaron. *Marconi's Wireless and the Rhetoric of a New Technology*. New York: Springer, 2012.

Varnava, Andrekos. 'Punch and the British Occupation of Cyprus in 1878.' *Byzantine and Modern Greek Studies* 29, no. 2 (2005): 167-186.

Weekly Dispatch. 'The Electric Light.' October 13, 1878.

Willis, Martin. *Vision, Science and Literature, 1870-1920: Ocular Horizons*. New York: Routledge, 2015.

2. The Far-Sight Machine and the Kinetograph

How Television Brought Liveness to the Cinema

Abstract

In 1889, Thomas Edison announced his invention of a 'Far-Sight Machine' in the lead-up to the Columbian Exposition (1893). When he unveiled his Kinetograph and Kinetoscope to the public in 1891, the subsequent confusion fuelled speculation about a hybrid electric-photography instrument that could transmit live images (like a television) as well as reproduce scenes (like the cinema). This discussion dovetailed into early cinema mythology that bolstered its identity as a spectacular attraction. This chapter explains how the 'Far-Sight Machine' transformed into the Kinetograph. The 'Far-Sight Machine' destabilizes assumptions about the distinctions between cinema and television, transmission and recording. It shows how the identities of cinema and television are deeply entangled with the social circumstances of their emergence.

Keywords: Thomas Edison; Columbian Exposition; Kinetograph and Kinetoscope; Cinema of Attractions

Thomas Edison's 1889 announcement that he intended to unveil a 'Far-Sight Machine' at the Columbian Exposition set the media ablaze.[1] A flurry of

1 See, for example, 'Edison's Last,' *Boston Journal*, 13 May 1889; 'A Far-Sight Machine,' *Electrical Review*, 25 May 1889; 'A Far-Sight Machine,' *Scientific American*, 1 June 1889; 'Untitled (Edison's Far-Sight Machine),' *Baltimore Herald*, 10 June 1889; 'What Next?' *Iron*, 14 June 1889; 'Fourth Edition' (Edison's Latest—A Far-Sight Machine), *Pall Mall Gazette*, 15 June 1889; 'Untitled' (Edison's Far-Sight Machine), *New York Graphic*, 26 June 1889; 'Untitled,' *The Illustrated London News* (American edition, NY), 19 August 1889; 'Mr. Edison and the Electric Millennium,' *Levant Herald*, 1 September 1889; 'The American Wizard: More Wonderful Things that Edison is to Bring Forth,' *Washington Post*, 1 September 1889; 'Edison's Talk,' *Brooklyn Journal*, 7 October

Roberts, I., *Visions of Electric Media: Television in the Victorian and Machine Ages*.
Taylor & Francis Group, 2019
DOI 10.5117/9789462986596_CH02

speculation ensued, consisting of a mixture of progressive rhetoric, satire, and scepticism on par with the 1878 discovery mania. Talk of a machine that would provide a visual accompaniment to the telephone fulfilled the expectation that the prophecy of 'Edison's Telephonoscope' had finally come true.[2] Meanwhile, Menlo Park technicians were ironing out the kinks of the Kinetograph and experimenting with ways to manufacture filmstrips in the laboratory. While film historians emphasize this period as the emergence of American cinema, journalists in 1889 were paying more attention to what Edison was saying about his new electric telescope, the 'Far-Sight Machine'.[3]

This enigmatic combination of photography and electricity was the source of much speculation. The *Electrical Review* reported: 'By means of this extraordinary invention he hopes to be able to increase the range of vision by hundreds of miles, so that, for instance, "a man in New York could see the features of his friend in Boston with as much ease as he could see a performance on stage."'[4] Commentary vacillated between the exaggerated hype of progress and the corrupting influence new technologies could have on wholesome American values. But without a functional model to demonstrate, newspapers lost interest in Edison's 'Far-Sight Machine' within a year.

The 'Far-Sight Machine' went down in history as a fantasy. The few mentions the machine has received in secondary literature in the history of film and media note the 'Far-Sight Machine' as a passing media spectacle.[5] The

1889; 'Mr Edison at Home Unspoiled by Glory,' *New York Herald*, 7 October 1889. See also Edison's newspaper clippings in his Far-Sight Machine, Menlo Park Scrapbook, 1889 (TAED SM035).

2 Thomas D. Lockwood, 'Observations,' *The Electrical Engineer* (US), October 1889, 423; 'Punch and the Kinetoscope,' *Illustrated American*, 20 June 1891, 224.

3 Tom Gunning, 'Doing for the Eye What the Phonograph Does for the Ear,' in *The Sounds of Early Cinema*, eds. Richard Abel and Rick Altman (Bloomington, IN: Indiana University Press, 2001); Gunning, 'An Aesthetic of Astonishment'; Tom Gunning, 'The Cinema of Attraction,' *Wide Angle* 3, no. 4 (1986); Tom Gunning, '"Primitive" Cinema: A Frame-up? Or the Trick's on Us,' *Cinema Journal* 28, no. 2 (1989): 3-12; Hendricks, *The Edison Motion Picture Myth* (Berkeley, CA: University of California Press, 1961); Gordon Hendricks, *The Kinetoscope: America's First Commercially Successful Motion Picture Exhibitor* (New York: Beginnings of the American Film, 1966); Charles Musser, *Before the Nickelodeon: Edwin S. Porter and the Edison Manufacturing Company* (Berkeley, CA: University of California Press, 1991); Charles Musser, *The Emergence of Cinema: The American screen to 1907* (Berkeley, CA: University of California Press, 1994); Paul Spehr, *The Man Who Made Movies: W.K.L. Dickson* (New Barnet, UK: John Libbey Publishing, 2008).

4 'A Far-Sight Machine,' *Electrical Review*, 25 May 1889.

5 Spehr, *The Man Who Made Movies*, 78, 174; Carolyn Marvin, 'The Electrical Imagination: Predicting the Future of Communications in Britain and the United States in the Late Nineteenth Century,' Diss., University of Illinois at Urbana-Champaign, 1979, 123; Stephen Herbert, *A History*

machine was never built. For all intents and purposes, it may as well have been a fiction. However, from a contextual perspective that acknowledges the role of the press, coverage of the 'Far-Sight Machine' played a crucial role in establishing expectations for early American cinema.

The press caught up with Edison again in May 1891 on his trip to inspect the Chicago site for the Columbian Exposition. In interviews, Edison continued to express his interest in 'seeing by electricity'. The character of his invention, however. had shifted slightly towards the photographic and mechanical. Coverage ceased mentioning the 'Far-Sight Machine' by name, instead referring to a 'happy combination of electricity and photography'.[6] Appeals to 'seeing by electricity' overlapped with the optical lantern, photography, and visual toys like the phenakistoscope and zootrope.[7] The 'happy combination of electricity and photography' exists at the intersection of cinema, media, and technology, suggesting a way of thinking about the moving image as a hybrid construction of social, technical, and economic forces. Edison's 'happy combination' performed many functions involved in both electrical transmission and photographic reproduction. Edison, along with the many journalists who contributed their own outlandish speculation, said that the invention would be able to reproduce entertainments as well as transmit live events. Edison said that, with his machine, 'a man can sit in his own parlor and see depicted on a curtain the forms of players in opera on a distant stage and hear the voices of the singers'.[8] Then he added: 'To the sporting fraternity I will state that ere long the system can be applied to prize fights [....] Arrangements can be made to send views of the mill ala stock and race ticker.'[9] Ramping up these speculations, a *Western Electrician* reporter later stated: 'It appears to be a device for reproducing photographs for moving objects on a screen at a distance from the scene portrayed at the time the event is transpiring or at a later date.'[10]

of Early Television (New York: Taylor & Francis, 2004), 3; Robert Conot, *A Streak of Luck: The Life and Times of Thomas A Edison* (Boston, MA: Da Capo Press, 1979), 346.

6 'Edison's Conjury,' *New York Sun*, 13 May 1891; 'The Kinetograph: Edison's Latest and Most Surprising Device,' *New York Sun*, 28 May 1891, 1-2. 'Happy combination' also refers to a statement in the Federalist Papers referring to the balance of state and federal power. Alexander Hamilton, John Jay, and James Madison, *The Federalist Papers* (The Floating Press, 2011), 72.

7 'A Move in the Right Direction – Mr. Edison's Latest,' *New York Times*, 29 May 1891; 'The Kinetograph,' *Electrical World*, 13 June 1891, 431; 'First Public Exhibition of Edison's Kinetograph,' *Scientific American*, 20 May 1893; 'Edison's Kinetograph,' *The Photographic News*, 16 June 1893.

8 'Edison's Conjury,' *New York Sun*, 13 May 1891.

9 'Edison's Conjury.'

10 'Edison's Visit to Chicago,' *Western Electrician* (Chicago), 23 May 1891, 295.

A month later, Edison changed his tune yet again. This time, Edison revealed that he would call his invention the Kinetograph. A series of demonstrations solidified this photographic-mechanical device as a real, functional machine for capturing 'living scenes'. Not long after, the Kinetoscope appeared, which would reproduce those scenes for a viewing audience. Press coverage hailed the Kinetograph and Kinetoscope as signs of Edison's *stick-to-it-iveness*. Journalists praised the inventor's mechanical genius. These inventions marked the dawn of the American cinema while effectively erasing the 'Far-Sight Machine' and the 'happy combination'. The new emphasis on photographic recording and reproduction eventually displaced any aspects of electricity or transmission in a practical sense. Paradoxically, however, the talk of the cinema retained a quality of liveness associated with electrical devices.

This chapter investigates announcements for and reception of the 'Far-Sight Machine', the 'happy combination of electricity', and the Kinetograph. Evidence from newspapers, magazines, and science journals shows how the electrical 'Far-Sight Machine' contributed to the early reception of and burgeoning identity of the Kinetograph and Kinetoscope, bringing the perception that cinema was imbued with 'liveness' along with it. The 'Far-Sight Machine' may have been doomed to failure, but it contributed to the success of the cinema in nineteenth-century America. These inventions signalled nineteenth-century expectations for seeing by electricity before categories of live television and recorded cinema coalesced.

Investigating Early Cinema Culture

By definition, film history focusses on the cinema once it has already emerged, taking care to mark distinct points in its trajectory towards a fully formed media concept. Tom Gunning and his long-time collaborator Andre Gaudreault define cinema culture as an object of study distinct from film texts, which emerged in America at the beginning of the 20th century.[11] They distinguish between cinema technology, cinematic representation, and the cultural experience of cinema. They look back into history in order to define that moment when cinema 'emerged into a form recognizable to us today. Writing of beginnings and emergence, their method investigates the cinema

11 Tom Gunning and Andre Gaudreault, 'Introduction: The American Cinema Emerges (1890-1909),' in *American Cinema, 1890-1909: Themes and Variations*, ed. Andre Gaudreault (New Brunswick, NJ: Rutgers, 2009), 1-21; Andre Gaudreault, *Film and Attraction*, 2-3.

and its culture as a distinct and identifiable object. Their periodization marks the birth of cinema in the 1890s and the emergence of cinematic practice, characterized by the appearance of nickelodeons (ramshackle storefront movie theaters) and trends towards commercialization after 1906.

Tom Gunning's 'cinema of attractions' remains the most widely acknowledged theory for comprehending early cinema.[12] In the article, Gunning compares silents from the nineteenth century with narrative films form the early 20th century. He introduces the term 'cinema of attractions' to identify the particularly exhibitionist entertainments of the 1890s. Juxtaposed with the narrative films that emerged after around 1906, the 'cinema of attraction' consisted predominantly of flashy spectacles and voyeuristic displays. Gunning's work succeeds in drawing similarities and differences between modes of cultural practice, such as vaudeville and fairground entertainments that contributed to the character of the 'cinema of attractions' as well as indicating the evolution of cinema from individual moments and standalone scenes to a classical style predicated on editing, continuity, and narrative logic.

Scholars of early cinema have devoted themselves to the study of inventions, films, and filmmakers, and the emergence of moving-image cultures. Investigations into the technological history of the cinema have included work on the Kinetograph and Kinetoscope, the manufacture of celluloid film, and the struggles to protect patent rights.[13] Others uncovered the development of cinema cultures in nickelodeons and Kinetoscope parlors, and later exhibitions and performances that incorporated films into their entertainment. Recent trends in social aspects of early cinema have led scholars to investigate evidence found in newspaper announcements and

12 Tom Gunning, 'The Cinema of Attraction.' See also: Wanda Strauven, ed. *The Cinema of Attractions Reloaded*; André Gaudreault, *Film and Attraction*.

13 Paul Spehr, 'Movies and the Kinetoscope,' in *American Cinema, 1890-1909: Themes and Variations*, ed. Andre Gaudreault (New Brunswick, NJ: Rutgers, 2009), 22-44; Brian Jacobson, 'The Black Maria: Film Studio, Film Technology,' *History and Technology* 27, no. 2 (2011): 233-241; Brian Jacobson, *Studios Before the System: Architecture, Technology, and the Emergence of Cinematic Space* (New York: Columbia University Press, 2015); Paul Moore, 'The Social Biograph: Newspapers as Archives of the Regional Mass Market for Movies,' in *Explorations in New Cinema History: Approaches and Case Studies*, eds. Richard Maltby, Daniel Biltereyst, and Philippe Meers (Hoboken, NJ: Wiley-Blackwell, 2011): 263-279; Santiago Hidalgo, 'Early American Film Publications: Film Consciousness, Self Consciousness,' in *A Companion to Early Cinema*, eds. André Gaudreault, Nicolas Dulac, and Santiago Hidalgo (New York: Wiley, 2012), 202-219; William Uricchio and Roberta Pearson, 'Coming to Terms with New York City's Moving Picture Operators, 1906–1913,' *The Moving Image: The Journal of the Association of Moving Image Archivists* 2, no. 2 (2002): 73-93.

other periodicals.[14] Gordon Hendricks was the first to dig into the archives to dispel myths that had been sustained since the birth of cinema, particularly having to do with the Kinetoscope. His landmark studies brought recognition to the work of W. K. L. Dickson, the Menlo Park technician who invented the Kinetoscope.[15] Hendricks used a combination of newspaper clippings and archival material to show how Edison took credit for Dickson's work. Since film historians focus on the reception of the Kinetoscope, little attention has been paid to the Kinetograph in particular. Once Dickson and Edison had introduced the Kinetoscope, popular discourse tended to fold these two inventions into one.[16] After 1895, mention of the Kinetoscope tended to refer generally to the cinematic apparatus consistent with the emphasis on exhibition and reproduction over the behind-the-scenes recording process.

This chapter takes advantage of scholarship that looks outside the traditional bounds of the film history. To think of the 'Far-Sight Machine' as an invention in the history of cinema and television raises the question of what distinguished an imaginary invention from a real, functional device. By showing how the *talk* of The 'Far-Sight Machine' played an important role in the development of the cinema, this chapter raises questions about the limitations of the categories for technologies and inventions. If the 'Far-Sight Machine' was truly an invention, it might best be described as an imaginary media artefact.[17] Because the machine never passed through the initial phase of speculation, the 'Far-Sight Machine' lives fossilized in the periodical record. All we have to rely on are words.

While film historians have studied the development of the Kinetograph and Kinetoscope in Edison's lab, insofar as it marks the beginning of an American film tradition, little has been written about the 'far-sight machine'. With no material artefact to speak of and no practical demonstrations of the device, the story amounts to the several times Edison mentioned his intention to build the device during interviews. Paul Spehr devotes a whole two paragraphs to the 'Far-Sight Machine' in the context of Edison's early work on the Kinetograph, treating it as a speculative, imaginary invention

14 Moore, 'The Social Biograph'; Hidalgo, 'Early American Film Publications'; Uricchio and Pearson, 'Coming to Terms with New York City's Moving Picture Operators.'

15 Hendricks, *The Edison Motion Picture Myth*, 104-105; Hendricks, *The Kinetoscope*.

16 W. K. L. Dickson and Antonia Dickson, *History of the Kinetograph, Kinetoscope, and Kinetophonograph* (Albert Bunn, 1895; New York: Museum of Modern Art, 2000).

17 Eric Kluitenberg, ed, *The Book of Imaginary Media: Excavating the Dream of the Ultimate Communication Medium* (Rotterdam, The Netherlands: NAi Publishers, 2007); Eric Kluitenberg, 'On the Archaeology of Imaginary Media,' in *Media Archaeology: Approaches, Applications, and Implications,* eds. Erkki Huhtamo and Jussi Parikka (Berkeley, CA: University of California Press, 2011), 48-69; Natale and Balbi, 'Media and the Imaginary in History.'

on par with the telephonoscope.[18] Spehr calls it mere 'fodder for journalists', a distraction from the real work of the Kinetograph. He relies heavily, as Hendricks did, on the archival documents and laboratory activities at Menlo Park. Additionally, Victorian visual-culture scholar Steven Herbert's article on the 'Far-Sight Machine' in the British humour magazine *Puck* remains one of the few scholarly examinations of the subject.[19]

Regardless of the method, film historians gravitate towards identifying the cinema as a distinct object of study. A contextual approach that identifies moving-image technologies as 'media in transition' broadens the discussion in order to consider the ways early cinema culture emerged from out of existing cultural practices. Just as the cultural circumstances preceding the publication of 'Edison's Telephonoscope' provide essential context, so too does the 'Far-Sight Machine' establish a tone for the reception of Edison's Kinetograph, Kinetoscope, and the later emergence of cinema. Understanding the Kinetograph and Kinetoscope as 'media in transition' supports a new way of thinking about the early cinema. Like Siegfried Zielinski's *Audiovisions* (1999) and William Uricchio's investigations into nineteenth-century film and television media identities, this chapter looks at a broader category that encompasses moving-image technologies to discover how the distinct media of television and cinema coalesced beginning in late nineteenth-century American culture.[20]

'Media in transition' has the advantage of locating the meaning of these emerging practices and technologies from the perspective of a contemporary observer or reader. Similar to the way the 'Telephonoscope' can seem to be a representation of television, the stories and myths about the early cinema make it seem to suggest it was inevitable to become an exhibition medium founded on photographic moving images. Locating the cinema in the historical moment when it transitioned from earlier cultural and technological forms dispels these assumptions. Letters to the editor of the London *Times* and the newspaper editorials printed in the New York *Sun* contextualize 'Edison's Telephonoscope'; similarly, the particular way of talking about photographic moving images, as it was represented in the mainstream American press around the time Edison introduced his Kinetograph and Kinetoscope, functioned to shape the meaning of the emerging media form.

18 Spehr, *The Man Who Made Movies.*

19 Stephen Herbert, 'Professor Goaheadison's latest,' *Early Popular Visual Culture* 9, no. 1 (2011): 75-81.

20 Siegfried Zielinski, *Audiovisions: cinema and television as entr'actes in history* (Amsterdam: Amsterdam University Press, 1999); Uricchio, 'Television, Film and the Struggle for Media Identity'; Uricchio, 'Storage, Simultaneity and the Media Technologies of Modernity.'

The 'Far-Sight Machine'

Edison announced his intention to build a 'far-sight machine' when approached with the question of what he would present at the next World's Fair. With the Paris Exposition Universelle in full swing, New Yorkers were reeling from a May Day ticker-tape parade in honour of the Washington Inaugural Centennial.[21] Exploiting the celebratory atmosphere, speculation grew on the possibility of a New York fair to be held on the anniversary of Columbus's arrival in America. Who better than Edison, the Menlo Park Wizard, a favourite on the exhibition circuit, to make a suggestion as to what could be expected from an 1892 Fair? P.T. Barnum also offered his suggestions, but a borrowed Egyptian sarcophagus would have paled in comparison to a magic mirror.[22] Whether identifying Edison's invention as the 'Far-Sight Machine' or just 'Edison's Latest,' this short notice, published in *Electrical Review* on 25 May 1889, spread like wildfire:

> Mr. Edison is reported, in a conversation with a reporter who solicited his ideas on the subject of the projected World's Fair in New York City, as saying that he would take an acre of space in such a fair and completely cover it with his inventions, of which he has no less than 70 now under way. 'One of the most peculiar and now promising good results,' said Mr. Edison, 'is what I may call a far-sight machine.' By means of this extraordinary invention he hopes to be able to increase the range of vision by hundreds of miles so that, for instance, a man in New York could see the features of his friend in Boston with as much ease as he could see a performance on stage. 'That,' he added, 'would be an invention worthy a prominent place in the World's Fair and I hope to have it perfected long before 1892.'[23]

21 'A World's Fair in 1892,' *New York Times*, 25 June 1889, 4. This article describes that the idea was first proposed during the 29-30 April centenary celebration of the inauguration of Washington.

22 P.T. Barnum, 'Notes and Comments: What the Fair Should Be,' *North American Review*, March 1890, 400-401.

23 'A Far-Sight Machine,' *Electrical Review*, 25 May 1889, 6; 'A Far-Sight Machine,' *Scientific American*, 1 June 1889; 'Untitled' (Edison's Far-Sight Machine), *The American* (Waterbury, CT), 22 May 1889; 'A Far-Sight Machine,' *The Statesman* (Yonkers, NY), 11 June 1889; 'Was Mr. Edison in Earnest?' *St. James's Gazette*, 12 June 1889; 'What Next?' *Iron*, 14 June 1889; 'Gleanings' (One of the most peculiar of Mr. Edison's recent inventions), *Birmingham Daily Post*, 15 June 1889; 'Edison's Far-Sight Machine,' *Western Mail* (UK), 15 June 1889; 'Edison's Coming Invention,' *Aberdeen Weekly Journal*, 15 June 1889; 'Edison's Latest Invention,' *Lloyd's Weekly Newspaper* (UK), 16 June 1889; 'Untitled' (Edison's Far-Sight Machine), *The Journal* (Pottsville, PA), 19 June 1889; 'A Far-Sight Machine,' *The Courier and London & Middlesex Counties Gazette*, 22 June 1889;

This notice curiously withholds details about how such a machine would be designed, how it would work, or what it might look like. The enthusiastic tone of the article leaves open a gap that journalists filled with their assumptions and expectations about the future of 'seeing by electricity'.

Public reception of the 'Far-Sight Machine' hinged on the enthusiasm for invention and the fantasies of scientific romance that filled the press. Journalist Horace Townsend recounts a visit to Menlo Park in an 1889 profile for *The Cosmopolitan*. Exploring Edison's library, he notes the plethora of speculative literature, a collection that encompasses both science fiction and newspaper articles proclaiming the invention of new devices.

> These articles were written and these statements signed but a few years ago, and to-day probably the very rooms in which they were penned are lighted by the incandescent filament enclosed in its airless bubble which has laid the foundations of its inventor's fortunes. Here too is a French novel, pasted into the book as it appears and its *feuilleton* form at the bottom of succeeding issues of the leading Parisian newspaper; and the marvelous hero of this blood-curdling romance, the scenes of which are laid in a New York possible only to the imagination of a French novelist, is Thomas A. Edison.[24]

Like Edison's nod to the science fiction in an inventor's imaginative life, Townsend takes note of the technological foundations of the scientific romance, probably referencing Villiers de l'Isle Adam or Albert Robida or Jules Verne. In many ways, the similarities between these genres are greater than their differences.

The similarities between Edison's 'far-sight machine' and the scientific romances of Verne, as well as Edward Bellamy's *Looking Backward*, both of which had been published just months before, were unmistakable. In early

'Untitled' (Edison's Far-Sight Machine), *The Time* (Bethlehem, PA), 25 June 1889; 'A Far-Sight Machine,' *The Post* (Liverpool, England), 17 June 1889; 'Untitled' (Edison's Far-Sight Machine), *The Times* (Portsmouth, NH), 29 July 1889; 'Untitled' (Edison's Far-Sight Machine), *The Leader* (Wilkes Barre, PA), 9 August 1889; 'Edison's Machines,' *Atchison Daily Champion* (Atchison, KS), 10 August 1889; 'Untitled' (Edison's Far-Sight Machine), *The Democrat* (Hutchinson, KS), 17 August 1889; 'The American Wizard: More Wonderful Things that Edison is to Bring Forth,' *Washington Post*, 1 September 1889; 'Mr. Edison and the Electric Millennium,' *Levant Herald*, 1 September 1889; 'The Past Year. A General Retrospective,' *Iron*, 3 January 1890; 'Engineering and Manufacturing Notes,' *American Engineer*, 7 May 1890; 'Notes: Electric Seeing,' *Electrical Engineer*, 30 May 1890; 'Chat,' *American Settler*, 12 July 1890. See also Edison's scrapbook of clippings: Far-Sight Machine, Menlo Park Scrapbook, 1889 (TAED SM035).

24 Horace Townsend, 'Edison: His Work and His Work-Shop,' *The Cosmopolitan*, 598-607: 602.

Figure 13. George Roux, 'Telephote,' in Jules Verne, 'In the Twenty-ninth Century: The Day of an American Journalist in 2889,' *Yesterday and Tomorrow*, 1910.

1889, Edward Bellamy's *Looking Backward* stole the spotlight in American circles for its political overtones.[25] At the same time, a short story attributed to Verne appeared in the American journal *Forum*. More so than Bellamy's Nationalist vision of the future, 'In the Year 2889' would have garnered attention from Edison and his fellow technology enthusiasts. Both novels explain how technology will serve essential social functions in the future. Bellamy describes how the 'electroscope' will fit into the fabric of his new society.

But Verne made no claims to a political agenda. Instead, the scientific romance told the story of a powerful Hearst-type mogul named Fritz Napoleon Smith, international newspaper magnate. The story is notable for the appearance of the *telephote*, a visual telephone (Figure 13) that broadcasts the news and weather while also enabling Smith to see his wife and children despite having to put in long hours at the office. As a technological

25 For other work on Bellamy and technology, see Verity Hunt, 'Electric leisure: late nineteenth-century dreams of remote viewing by "Telectroscope",' *Journal of Literature and Science* 7, no. 1 (2014): 55-76; Howard Segal, *Technological Utopianism in American Culture* (Syracuse, NY: Syracuse University Press, 1985); Howard Segal, *Future Imperfect: The Mixed Blessings of Technology in America* (Amherst, MA: University of Massachusetts Press, 1994).

connection between Smith's office and his home, the *telephote* functioned as more than just a communication medium. The story framed the device as a way to maintain an intimate familial bond despite the separation of space. *Electrical Review* summed up the American reaction to Verne's futuristic vision:

> The editor [Smith] rules the world [...] He not only has a telephone line to Paris but a *telephote* as well, whereby he can at any time, from his study in New York, see a Parisian with whom he converses [....] Reporters describe events orally to millions of subscribers; and if a subscriber becomes weary or busy, he attaches his phonograph to his telephone, and hears the news at his leisure.[26]

The story summary highlights the uses of the futuristic technology in a way similar to how popular science speculates on the practical applications for everyday life. Appealing to the imagination of inventors and readers alike, the *telephote* functions in the story as more than a visual adjunct to the telephone. Depicted as a tool for the elite newspaper editor, it exaggerates the power of those who already possess it. It would have been different, for example, if the novel had dramatized the *telephote* conversation of the Sri Lankan nursemaid depicted in Punch's 'Telephonoscope'. Verne and Edison implicitly established a technological divide, suggesting a race, class, and gender for the ideal *telephote* user. If 'seeing by electricity' was going to make the world a better place, it would start at the top and trickle its way down.

One caveat, the preliminary patent application filed by Edison that May, reveals the kind of design he might have had in mind for the 'far-sight machine'. Labelled a device for 'telegraphic photography', the description makes the proposed invention sound like an electrical device for picture transmission. The name recalls the work of electrical engineer Shelford Bidwell, who had gained prominence in England for his work on tele-photography. Bidwell's method for the telegraphic transmission of still photographs followed a line of engineers who, since the early years of telegraphy in the 1840s, had struggled to refine the mechanics and synchronization to scan and reconstitute a picture line by line, as indicated in Figure 14.[27] Edison's designation of his device, using the name 'telegraphic photography', associated his invention with the methods of

26 'How Electricity Will Help the Editor of the Future,' *Electrical Review*, 2 February 1889, 4.

27 The *Brooklyn Journal* indicated a correspondence between Edison's 'Far-Sight Machine' and Bidwell's telegraphic photography in: 'Edison's Talk,' 7 October 1889.

Figure 14. Shelford Bidwell, 'Telephotography,' *Nature* 10 February 1881, 344-346.

facsimile reproduction. But the visual depiction of the device resembles something different altogether (Figure 15). Edison describes the system as a way of seeing at a distance optically: 'Long tubes [...] connected together and made airtight', and adjustable 'prisms' so that 'the curvature of the earth is corrected'.[28] 'A brilliantly illuminated object situated at one end may be perceived at the other end many miles distant.' Mirrors placed in a vacuum tube refract the light bouncing off distant objects in order to present the viewer with an image of the object as if through a window. The appearance of the design suggests a device more closely related to 'seeing by electricity' than to 'telegraphic photography'.

Journalists frequently made reference to 'far-sight' as a play on words. 'Far-sighted' and 'sagacious' appear together frequently in late nineteenth-century periodicals. British humour magazine *Fun* articulated the metaphor best as a play of sight and wisdom. 'A *short* sighted person requires a *high* glass; a *far* sighted person a *nigh* glass.'[29] Just as sight allows one to navigate space, wisdom allows one to navigate time. *Fun*'s aphorism suggests a distinct televisual culture of time, space, and mind. 'Far-sight' was also a character trait, as in the case of French chemist Henri Courtonne. Making press in France in competition with Edison, they called him a 'far-sighted individual' as a play on both the ability to affect the technological future and the power to see into the distance.[30] 'Far-sight' functioned with two integral connotations in nineteenth-century rhetoric. The *American Dictionary* (1897) lists two different uses of the phrase, one literal and the other figurative.

28 Patent Series PT031AAF1: 26, 59 (TAED Case 115).

29 'The Long and Short of It,' *Fun*, 19 June 1878. See also *Judy*, 'A Run on "Far-Sight Machines",' 26 June 1889; 'Untitled' (Edison's Far-Sight Machine), *The World* (Omaha, NE), 24 May 1889.

30 'A Far-Sighted Individual,' *Electrical Review*, 2 August 1889, 126. For more press coverage of Courtonne, see 'The Telephote,' *Telegraphic Journal and Electrical Review*, 16 August 1889; 'Untitled' (Edison's Far-Sight Machine), *The Beacon* (Akron, OH), 10 August 1889; 'Untitled,' *The Illustrated London News* (American edition, NY), 19 August 1889.

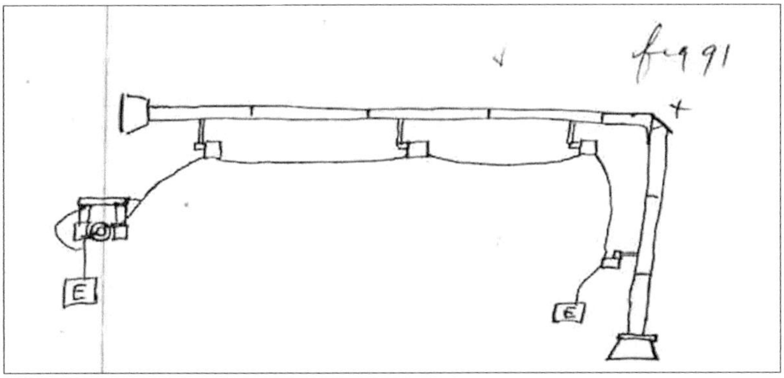

Figure 15. Thomas Alva Edison, 'Telegraphic Photography,' Patent Series — Caveat Files: Case 115: Electric Light, Ore Milling, Phonograph, Telegraph, Telephone (12 May 1889) [PT031AAF1; TAEM 113:497].

Literally, 'far-sight' meant to see at a distance, as either a quality of the eye or a relationship between the individual and the distant horizon. In this sense, 'far-sight' refers to a cultural impression of *space*. In the second, figurative sense, it was used in much the same way we use the word 'foresight' today. In this sense, seeing is linked with *time*.

The initial announcements that appeared in the press praised Edison for his genius invention. The press was saturated with exaggerated claims about technological progress, the dawn of a new age, and the power of machines to change the world. A reporter for the *West Chester Record* (PA) stated: 'Of the many wonders electricity has in store for mankind, we have probably no adequate conception of what the next quarter of the century will bring forth.'[31] Speculation swayed the other way as well. Talk of the 'Far-Sight Machine' also attracted scepticism from journalists who could imagine how such a device would *not* make the world a better place. The Pall Mall Gazette proffered: 'Query, will the farsight machine add to the joy or the misery of the world?'[32]

Responses also came dripping with sarcasm. Critics sceptical of Edison's claims attacked the inventor for stretching the truth. Lockwood, of the *Electrical Engineer*, put it eloquently: 'But much of this seems indeed to have a very

31 'Untitled' (Edison's Far-Sight Machine), *The Record* (West Chester, PA), 18 July 1889. See also 'Remnants – Some Wonderful Inventions,' *Detroit Free Press*, 6 October 1889; 'Untitled' (Edison's Far-Sight Machine), *The Commercial* (Toledo, OH), 12 August 1889; 'Untitled' (Edison's Far-Sight Machine), *Saint Paul Daily Globe*, 6 February 1890; 'Edison's Latest,' *Sacramento Daily Record*, 3 February 1890.

32 'Fourth Edition' (Edison's Latest—A Far-Sight Machine), *Pall Mall Gazette*, 15 June 1889.

ancient and fishlike smell.'[33] Edison's talk of a 'Far-Sight Machine' became the butt of endless jokes, the target of rumours, and the source of unrestrained speculation in 1889. Apart from the many reprinted articles from popular science journals such as *Electrical Review* and *Iron*, the bulk of the media coverage waxed sceptical.[34] Not everybody was convinced that the 'Far-Sight machine' would make the world a better place. In stark contrast to the usually enthusiastic hype of Edisonian invention, reactions to the 'far-sight machine' resemble those of the critics of the phonograph when it first appeared on the scene in 1878. Commentary published in the *Boston Journal* in early May established many of the themes that would play out in later satires: fears and anxieties about technological change, particularly as it concerned changes to social customs and privacy. Mixed in with the exaggerated announcement of 'Edison's Latest', the reporter offers speculations related to how the 'Far-Sight Machine' would have a detrimental effect on society:

> What a changed world it would create. It would be no more pleasant little deceptions about 'being at the club until the late hour.' The suspicious wife, supplied with the invention, would consult it, and would triumphantly prove that the husband was not at the club. The 'electric mirror,' by enabling persons hundreds of miles away to have whole tracts, whole city districts, whole categories of individuals under their examination, would place a new and wholesome restraint upon human action. Hypocrisy would go out of fashion, because it would be no longer practiceable; in fact, it would be extremely dangerous. Society would have to exist in unison with truth. Private detectives would give up their profession and go for something respectable. A 'fierce white light' would beat around the actions of every individual. But Edison would be burned, or something worse, for witchcraft. There will be a panic, and the old barbaric rage against the innovation would break forth. Let him be warned in time and place this invention on the shelf, labeled 'dangerous: too revolutionary.[35]

33 Thomas D. Lockwood, 'Observations,' *The Electrical Engineer* (US), October 1889, 423. See also 'Untitled' (Edison's Far-Sight Machine), *Freeman's Journal and Daily Commercial Advertizer* (Dublin, Ireland), 21 June 1889; 'Untitled' (Edison's Far-Sight Machine), *The Washington Critic*, 11 February 1890

34 See, for example, 'Grey's Vanishing Room,' *Chicago Daily Tribune*, 31 July 1890; 'Personal and Impersonal' (If Edison had his Far-Sight machine in order), *Milwaukee Daily Journal* (Milwaukee, WI), 24 May 1889; 'The Wizard of Menlo Park,' *The Ripley Journal* (Osgood, Indiana), 17 April 1890; 'Wizard of Menlo Park,' *Milwaukee Daily Journal* (Milwaukee, WI), 26 April 1890; 'Gossip on the Quiet,' *Yenowine's News* (Milwaukee, WI), 23 June 1889.

35 'Edison's Last,' *Boston Journal*, 13 May 1889.

The *Journal*'s announcement set the tone for the criticism that followed. Many journalists rejected the machine because of the consequences it could have for privacy. Some also positioned the 'Far-Sight Machine' as an indicator of the negative role of technology in changing social customs. According to the *Journal*'s radically sarcastic comment, for example, 'society would have to exist in unison with truth'.

Punch uses wit and humour instead of sarcasm. 'Open House (To be dated after the next invention)' narrates the conversation of a 'Far-Sight Machine' user gazing on his neighbours and family members from afar. The article pokes fun that one's facial expressions can belie the words written in innocent thank-you letters. The 'Far-Sight Machine' would reveal that hypocrisy.[36] Just as the 'Open House' hinted at the discrepancy between words and actions, *Punch* further dramatized the aspects of surveillance made possible by telephones, telephonoscopes, and the 'Far-sight Machine'. 'What it Might Come to In London' satirizes the exaggerated expectations that had grown around these new technologies. In it, a 'lecturer' demonstrates Edison's novel devices: the kinetograph, phonograph, and telephone. A 'mild young lady' comes forward hoping to see a recently married friend on her honeymoon only to discover, to her dismay, the couple in the midst of a heated argument. This article plays out the consequences for the surveillance after the 'annihilation of space'.

Judy's satire of the 'Far-Sight Machine' also highlights the effect the machine might have on privacy. Mrs. Penhecker 'is anxious to keep an eye on her husband's horrid club the next time Mr. Penhecker is detained in the city.'[37] Another character hopes that 'the apparatus will enable a man to place his neighbors [...] under constant supervision'. Other papers chimed in, suggesting that the 'Far-Sight Machine' would be a useful tool for wives to spy on their cheating husbands.[38] The London *Penny Illustrated* offers a rare visualization of the 'far-sight machine' in 'Henley and the March of Science' (Figure 20). The fixture, depicted here as a periscope-like peephole, was left open, allowing the young wife to catch a surreptitious glimpse of her husband's extracurricular activities.

The 'Far-Sight Machine' opened the floodgates to the possibility that anybody could not only show up without a moment's notice, but also show

36 'Open House (To be dated after the next invention),' *Punch*, 10 August 1889.

37 'A Run on Far-Sight Machines,' *Judy*, 26 June 1889. See also 'Pepper and Salt,' *Judy*, 26 June 1889; 'The Twigger,' *Judy*, 26 June 1889.

38 'Untitled (Edison's Far-Sight Machine), *Republican* (Meridan, CT), 14 May 1889; 'Untitled' (Edison's Far-Sight Machine), *Commercial Gazette* (Cincinnati, OH), 21 July 1889.

Figure 16. 'Henley and the March of Science,' *Penny Illustrated* (London), 6 July 1889.

their face. Several journalists expressed the fear of instant intrusions by in-laws and tax collectors: 'The new invention may not be so pleasant when you are "rung up" by the fellow whose bill is a little past due!'[39] Several newspapers also spouted jokes that the 'Far-Sight Machine' would add an undesirable face to the telephone exchange. The *Baltimore Herald* complained that '[t]he public will not only be treated to vocal wrath, as now, when an irritable patron is calling up "Exchange," but I will be enabled to see the violent contortions of his physiognomy in his attempt to secure the correct number'.[40]

The *Portland Commercial* added that the Far-Sight Machine would catch the telephone off-guard:

> A telephonic 'Be mine!' or 'Name the day' might, with nice discretion and tact, be answered by a suddenly-flashed vision of downcast eyes, trembling lashes, and 'a maiden blush' tingling a smile, as it were the roseate hues of early dawn caught in the silver rippling of a stream. The tele–smile or the tell–tale–tele–blush would come in useful here, and per contra, so

39 'Untitled' (Edison's Far-Sight Machine), *Tribune* (Racine, OH), 24 July 1889. See also 'What the Boys Are Saying,' *The Evening Tribune* (Lawrence, KS), 4 June 1889; 'Untitled' (Edison's Far-Sight Machine), *The Journal* (Lincoln, NE), 26 May 1889.

40 'Untitled' (Edison's Far-Sight Machine), *Baltimore Herald*, 10 June 1889.

> would be tele-frown of the indignant father, who would simultaneously shout through a megaphone, 'I am astonished at your imprudence, sir!'[41]

Characterized by fears of the ways technology was poised to change social standards, the criticism that erupted centred on the invasion of privacy and the betrayal of secrets. The 'far-sight machine' suggested a completely new social configuration which would make it possible for privileged users to experience the drama from the exclusive location of their private dwelling.

The British humour magazine *Fun* sums up the speculation surrounding the 'Far-Sight Machine' in 'Goaheadison's Latest' (Figure 17a). Read the title aloud: a mashup of go-aheadism and Edison.[42] This illustration of the

41 'Untitled' (Edison's Far-Sight Machine), *The Commercial* (Portland, IN), 5 September 1889.
42 Scott Sandage, *Born Losers: A history of failure in America* (Cambridge, MA: Harvard University

Figure 17a. 'Professor Goaheadison's Latest,' *Fun*, 3 July 1889, 6.

Press, 2005), 26-27; 'Professor Goaheadison's Latest,' *Fun*, 3 July 1889, 6; 'Goaheadison's Real

Figure 17b. 'Goaheadison's Real Latest,' *Fun*, 17 July 1889, 24.

'Far-Sight Machine' depicts a visual telephone capable of bringing patients in closer contact with their doctors and the forceful blows of boxing matches straight into the home (Figure 17b). *Fun*'s mashup of go-aheadism with the name of Edison signalled an unwillingness on the part of the inventor to back down from the pressure of technological progress. This satire indicates that, on the other side of the debate exists a culture unwilling to accept radically new technologies. *Fun* asks: will such a machine actually make the world a better place, or merely more of a spectacle?

Journalists also rejected Edison's 'Far-Sight Machine' in fear that it could radically change social customs, particularly the role played by audiences in participating in the entertainment as well as the social scene. The 'far-sight machine' held the potential to antiquate the act of going to the theatre altogether. The most common character cited as evidence of these social changes was known as the 'man who goes out between the acts'. As stories circulated about its potential uses, journalists often joked that the only person who would want to use the machine would be 'the man who goes out between the acts', the chaperone who leaves his date at intermission presumably in order to smoke cigars and drink in the lobby.[43] Associating his behaviour with the 'far-sight machine' suggested that the new technology would only encourage similar lapses in judgment.

The tone of sarcasm and satire characteristic in criticism lodged against 'Far-Sight Machine' was also prevalent in responses to Edison's phonograph.

Latest,' *Fun*, 17 July 1889, 24; Stephen Herbert, 'Professor Goaheadison's Latest,' *Early Popular Visual Culture* 9, no. 1 (2011): 75-81.

43 'What Edison Claims,' *Chicago Journal*, 13 May 1891. The characteristic was described at length as follows: 'A gentleman should on no account leave the lady's side from the beginning to the close of the performance. The custom of going out alone between the acts to visit the refreshment room cannot be too strongly reprehended. It is little less than an insult to the lady.' 'Between the Acts,' *Illustrated American*, 4 October 1890, 3.

Back in 1878, popular response to the power of the phonograph to capture fugitive sounds for posterity intersected with the character of a would-be user. He was identified as 'irreverent' as early as August 1878:

> As for the phonograph, his [Edison's] faith in it is boundless. In future, he believes, letters will be talked, books read, sermons preached, languages and music taught, parlor operas played, announcements made, and reporting done by phonograph. Voice-albums will become the fashion, and the memorable words of great men will be treasured in museums. 'There was a fortune in the Pope's last blessing,' says Edison, somewhat irreverently; 'the phonograph record of it, multiplied by electrotyping, would have sold for five dollars a piece easily.'[44]

The 'irreverence' of new technology became a trope in popular culture that emphasized its negative influence on social change. The power of the phonograph to capture fugitive sounds came across to some as an unnatural ability. A machine that would make possible the presence of speakers in their bodily absence corrupted the natural order of things.

Rejection of the inhuman abilities of the phonograph resurfaced during the promotion of the 'far-sight machine', which overlapped with Edison's repackaged 'talking phonograph' and 'talking doll' (Figure 18). Differentiated from the invention that made him famous in 1878, Edison unveiled a machine that could do much more than record speech. It stood in for the voice. Emphasis shifted from *recording* to *reproduction*. The new talking phonograph could now simulate speech, making it seem like the speaker was present in the room.[45] According to the London *Pall Mall Gazette*, the 'talking phonograph [...] added a new horror to existence'.[46] American criticism of phonographic recording had a much more whimsical, bemused tone. Journalists mused on the potential for high jinks. For example, the story about Edison's new mechanical stand-in made national news:

> Edison was unable to make it to the Electric Light Convention in Kansas City, so he sent a phonograph, and the little machine delivered an address

44 'Edison, The Inventor of the Phonograph,' *Engineering and Mining Journal*, August 1878, 131.

45 James Lastra, *Sound Technology and the American Cinema: Perception, Representation, Modernity* (New York: Columbia University Press, 2000), 21-22.

46 'Untitled' (Edison has added a new horror to existence), *Pall Mall Gazette* (London), 3 February 1890. See also 'Untitled' (With all its 'hem's and 'ha's), *Morning Advertiser* (London), 4 February 1890; Litigation Records Series, Thomas A. Edison v. American Mutoscope Company, 1898-1900 (TAED QM001).

Figure 18. 'Edison's Phonographic Doll,' *Scientific American*, 26 April 1890.

> for him, which greatly pleased the members of the association. Pretty soon the business will be done so fine that people, instead of attending evening parties, will just charge their phonograph, and send it by a servant. What a fearful clatter of gossip there would be if all the machines would go off at once! And yet it often happens that a great many mouths get to work at once, even as it now is.[47]

This story raised questions as to the social utility and implications of recording in a joking manner. But not all responses were so jovial. One story about a recently deceased lunatic's ravings recorded on a phonograph was enough to raise the hair on the back of one's neck.[48] All in all, the stories about the 'irreverence' of the phonograph and Kinetograph began circulating in early 1890 related to fears of the potential implications of recording technology for physical presence, which should come as no profound surprise. The phonograph ushered in an era in which the possibility of emotive presence after

47 'Untitled' (Edison at the Electric Light Convention), *Perrysberg Journal*, 15 February 1890.

48 'Untitled' (One of the curiosities of Edison), *News and Citizen* (Morrisville, VT), 13 February 1890; 'A Dead Madman's Words,' *Wahpeton Times* (Wahpeton, ND), 13 February 1890.

death was but a push of a button away. The 'irreverence' of the phonograph referred to the disrespect for the lived, ephemeral moment.

'A Happy Combination of Electricity and Photography'

By the end of 1889, Edison's 'Far-Sight Machine' had gained a reputation similar to the telephonoscope as a result of the mass of newspaper articles and satires declaring the machine's transgression of established social customs. At the same time that speculation had been circulating about Edison's 'Far-Sight Machine', news began trickling out of Menlo Park that several technological advances could make it possible to photograph a speeding bullet and to capture a speaker's gestures like the phonograph caught speech. The *Herald* announced the invention of 'a photographic adjunct to the phonograph', to which the Atlanta Constitution responded, 'how thankful should we be' if Edison's latest invention could 'catching speaker's gestures' in Congress.[49] These articles promoted advances in 'instantaneous photography' by emphasizing the 'realistic picture' reproduced.[50]

Journalists speculated on the photographic-mechanical machine in much the same way Edison had offered possible applications for the 'Far-Sight Machine'. While the *Herald* promoted Edison's progress in 'instantaneous photography', the *Sun* insisted that Edison's latest would indeed live up to the dream of 'seeing by electricity'.[51] Combining the ideas of both machines, the *Sun* declared that Edison was working on a device for 'seeing by electricity' that would integrate the developments seen in the field of photography. In 'Electric Marvels', the *Oregonian* emphasizes the mechanical nature of Edison's latest invention: 'It is said to be possible that modern electricians may succeed in constructing a device that will do for the sense of sight what the telephone does for the sense of hearing; but the prospects of such an achievement are not particularly bright.'[52] Continuing along this vein of speculation, a Kentucky paper claimed that Edison's latest 'will transmit and

49 'To Catch a Speaker's Gestures,' *New York Herald*, 2 February 1890; 'Latest from the Wizard,' *Sacramento Daily Record*, 3 February 1890; 'People Here and There,' *The Atlanta Constitution* (GA), 14 February 1890.

50 'To Catch a Speaker's Gestures,' *New York Herald*, 2 February 1890; 'Untitled' (Edison has added a new horror to existence), *Pall Mall Gazette* (London), 3 February 1890.

51 'To See By Electricity: An Apparatus for the Eye as the Phonograph is for the Ear,' *New York Sun*, 12 September 1890.

52 'Electric Marvels: An Apparatus for the eye as the phonograph is to the ear,' *Morning Oregonian*, 16 November 1890.

reproduce motion of any kind for any distance'.[53] These articles effectively merged the 'Far-Sight Machine' with 'instantaneous photography'.

The story resurfaced a year later. In May 1891, Edison travelled to Chicago to inspect the site of the Electrical Exhibit. Journalists flocked to hear news of his inventions. Several interviews published in May unveil what Edison then called the 'happy combination of electricity and photography', described alternatively as a device for seeing by electricity and a photographic adjunct to the phonograph. The 'happy combination' retained the novel characteristics of the 'Far-Sight Machine' while integrating the developments his laboratory personnel had achieved in 'instantaneous photography'. The announcements that came out in May of 1891 made Edison's latest invention out to be a combination of photography, phonography, and electricity.

In the absence of a material device, demonstrations, or even descriptions of how it would work, journalists struggled to understand the machine. One version emphasized the machine as a photographic-mechanical adjunct to the phonograph. Descriptions of this type took note of the phrase 'to do for the eye what the phonograph does for the ear', which was repeated from Edison's 1888 patent application for the Kinetograph.[54] This version appealed to the reproduction of operas and boxing fights, the prospect of home entertainment, and the fidelity of the image to the actual scene.

The *Chicago Evening Post* announced:

> When this invention shall have been perfected said Mr. Edison with the trace of enthusiasm's glow in his face, a man will be able to sit in his library at home and having electrical connection with the theatre, see reproduced on his wall or a piece of canvas the actors and hear anything they say.[55]

53 'The Kinetograph: With it You can See a Man a Thousand Miles Away—Edison's Latest,' *The Big Sandy News* (Louisa, KY), 28 April 1890.

54 'Edison's in Chicago: The Wizard of Menlo Park Stopping at the Auditorium, Tells of his Latest Invention,' *Chicago Evening Post*, 12 May 1891; 'Edison's Latest,' *Daily Citizen* (Asheville, NC), 28 May 1891; 'Edison and the Big Fair,' *Chicago Tribune*, 14 May 1891; 'Edison's Visit to Chicago,' *Western Electrician* (Chicago, IL), 23 May 1891, 295; 'Untitled' (Edison's promised Kinetograph), *Daily Intelligencer* (Wheeling, WV), 25 May 1891; 'The Kinetograph,' *Evening World* (New York), 28 May 1891; 'The Month – Science and Arts,' *Chambers' Journal*, 25 July 1891, 477; 'Edison's Latest,' *American Journal of Photography* (reprinted from the *Philadelphia Daily News*), June 1891, 338; 'Light Wedded to Sound,' *Indianapolis Herald*, 31 May 1891; 'The Kinetograph,' *Phonogram*, October 1892, 217-219; 'First Public Exhibition of Edison's Kinetograph,' *Scientific American*, 20 May 1893; 'Edison's Kinetograph,' *The Photographic News*, 16 June 1893. These articles emphasize the Kinetograph as a combination of photography and phonography.

55 'Edison's in Chicago: The Wizard of Menlo Park Stopping at the Auditorium, Tells of his Latest Invention,' *Chicago Evening Post*, 12 May 1891; 'Edison's Latest,' *Daily Citizen* (Asheville, NC), 28 May 1891.

The *Chicago Tribune* focussed on Edison's mastery of reproduction: 'When the process is completed, the reproduction will be lifelike.'[56] Such statements introduced the combination of photography and phonography as a device that could reproduce a lifelike copy with precision and fidelity. Discussions of 'instantaneous photography' emphasize Edison's mastery in reproducing a lifelike, realistic image.[57] As one correspondent for the *Pittsburg Dispatch* explained, 'the result is that the eye does not see the forty-six photographs, but it sees only the one with the motions or gestures of the man taken'.[58] Calling the filmstrip a 'movement record', journalist Lathrop for *Harper's Weekly* eloquently wrote that the Kinetograph 'reproduces with absolute fidelity and naturalness the *movement* as well as the form of the original object'.[59] *Illustrated American* noted how recordings of 'actual events' could be 'presented to the eye and ear with the fidelity of life'.[60] Talk of recording always emphasized the 'fidelity and naturalness' of the reproduction, creating a lifelike impression.

The New York *Sun* and London *Times* also ran cables from correspondents that put a spin on the photographic reproduction. These articles introduced another interview with Edison that identified his new invention alternatively as a 'happy combination of photography and electricity' and a 'Photophone Kinetograph'.

> My intention is to have such a *happy combination of photography and electricity* that a man can sit in his own parlor and see depicted on a curtain the forms of players in opera on a distant stage and hear the voices of the singers. When the system is perfected, which will be in time for the fair, each little muscle of the singer's face will be seen to work, every color of his attire will be exactly reproduced and the stride and positions will be natural and will vary as do those of the person himself. To the

56 'Edison and the Big Fair,' *Chicago Tribune*, 14 May 1891; 'Edison's Visit to Chicago,' *Western Electrician* (Chicago, IL), 23 May 1891, 295; 'Untitled' (Edison's promised Kinetograph), *Wheeling Daily Intelligencer* (Wheeling, WV), 25 May 1891; 'The Kinetograph,' *The Evening World* (New York), 28 May 1891; 'The Month – Science and Arts,' *Chambers' Journal*, 25 July 1891, 477.

57 'Occasional Notes,' *Pall Mall Gazette* (London), 29 May 1891; 'Science and Invention: Edison's Photophone Kinetograph,' *The Newcastle Weekly Courant*, 13 June 1891; 'Edison's Photophonokinetograph,' *Electrical Engineer*, 20 May 1891, 584; 'Edison's Photophonokinetograph,' *English Mechanic*, 20 May 1891.

58 Frank Carpenter, 'Edison in his Den,' *The Pittsburg Dispatch*, 5 November 1891. See also Oscar Davis, 'Edison in his Laboratory,' *Electricity*, 22 July 1891, 5.

59 George Parsons Lathrop, 'Edison's Kinetograph,' *Harpers Weekly*, 13 June 1891, 446-447; 'The Edison Kinetograph,' *The Electrical Engineer*, 24 June 1891, 708.

60 'Mr. Edison's Kinetograph,' *Illustrated American*, 20 June 1891, 194.

> sporting fraternity I will state that ere long the system can be applied to prize fights. The whole scene, with the noise of the blows, talk, etc., will be truthfully transferred. Arrangements can be made to send views of the mill ala stock and race ticker.[61]

In spite of the emphasis these May announcements placed on the photographic and mechanical (recording) aspects of the invention, reporters continued to promote the machine as an electrical device. Announcements persisted in describing the invention as a combination of electricity, photography, and phonography, explaining the many possible uses of the device for the reproduction of scenes *both live and recorded.*

The New York *Sun*'s front-page announcement of Edison's invention of the Kinetograph gave his combination of electricity, photography, and phonography a distinct name, an associated illustration (Figure 19), and a reinvigorated story.[62] In the characteristically exaggerated tone of the *Sun*, the journalist unveiled the Kinetograph as a sign of Edison's triumph over adversity:

> Three or four years ago, in a magazine article, Edison, the electric wizard, wrote that he would produce a machine which should record and reproduce motion as the phonograph recorded and reproduced sound. Other electrical would-be wizards pooh-poohed the scheme. The electrical periodicals scouted the idea, and *irreverent newspapers* told Edison that he talked too much[....] People laughed for a while but Edison kept still and they forgot the Wizard's 'wild scheme.' [...] He worked for more than three years, and at last was successful, in so far as to correctly establish his 'germ' or 'base principle.' Then Edison laughed. He sat in the big armchair in his laboratory and watched a crude model of the machine and thought of what a lot of fun he would have with the people who had told him he talked too much.[63]

61 'Edison's Conjury,'*New York Sun*, 13 May 1891; 'Mr. Edison and the Chicago Exhibition,' *Times* (London), 14 May 1891; 'Edison's Photophonokinetograph,' *Electrical Engineer*, 20 May 1891, 584; 'Edison's Photophonokinetograph,' *English Mechanic*, 20 May 1891; 'Edison's Kinetograph,' *English Mechanic*, 5 June 1891, 310; 'Mr. Edison's Latest Invention,' *Times* (London), 29 May 1891; 'Mr. Edison's Latest Invention,' *Pall Mall Gazette* (London), 29 May 1891; 'Edison's Kinetograph,' *Manufacturer and Builder*, June 1891, 138; 'The Advance of Science,' *Iron*, 5 June 1891.

62 'The Kinetograph', *New York Sun*, May 28, 1891; 'Edison's Kinetograph,' *Scientific American*, June 20, 1891, 393; George Scott, 'Edison's Latest Marvel: The Kinetograph,' letter to the editor, *English Mechanic*, June 19, 1891, 358-359.

63 'The Kinetograph: Edison's latest and most surprising device,' *New York Sun*, May 28 1891.

Figure 19. 'The Kinetograph,' *The Sun*, 28 May 1891, 1-2.

Having already reviewed the lead-up to this announcement, we can begin to untangle the *Sun*'s outrageous claims. In this article, the *Sun* journalist rearticulates the criticism lodged against the phonograph's 'irreverence' and instead positions this criticism as an advantage. The story frames Edison as a success: an inventor who has surmounted the obstacles and succeeded against adversity. For all intents and purposes, the *Sun* pictures Edison in exactly the image that *Fun* had satirized years ago: Professor Goaheadison.

Also recall that the Kinetoscope had garnered negligible response when Edison mentioned it to the press in 1890.[64] But, juxtaposed with the critical

64 Horace Townsend, 'Edison: His Work and His workshop,' *The Cosmopolitan: A Monthly Illsutrated Magazine*, April 1889; 'To Catch a Speaker's Gestures,' *New York Herald*, 2 February 1890; 'Latest from the Wizard,' *Sacramento Daily Record*, 3 February 1890; 'People Here and There,' *The Atlanta Constitution* (GA), 14 February 1890.

backlash produced against the phonograph and the 'far-sight machine', Edison's 'vow' makes much more sense. Intentionally drawing on these earlier indictments, the *Sun* reaffirms Edison's tenacity. Having vowed to succeed against all odds, the inventor now had something to show for all his efforts.

Towards the end of the article, the *Sun* reporter finally addresses the elephant in the room: 'At the first blush it seems to be that Mr. Edison has found a scheme telegraphing the representation of action. This not the case.'[65] Despite this disclaimer, journalists continued to speculate on the electrical nature of the kinetograph. In 'Punch and the Kinetograph', *Illustrated American* hailed Edison's Kinetoscope as the fulfilment of a prophecy foretold over a decade earlier by British illustrator George Du Maurier. It should come as no surprise that minds jumped to the 'telephonoscope' upon the announcement of a similar invention. Du Maurier's illustration pictured an imaginary invention of Edison's, intended as a visual adjunct to the telephone that would make it possible for parents in London to hear as well as see their daughter living on the other side of the world. 'The happy combination of electricity and photography' seemed to do just that. Specific references to British culture and the electric-light mania were lost to an American audience who, reading it in the context of the 1891 unveiling of the Kinetograph, saw only progress. The magazine declared the Kinetoscope the achievement of the age: 'The world ceases to scoff. It marvels.'[66]

Additionally, *Engineering* and *Leisure Hour* both announced the Kinetograph with reference to 'seeing by electricity'. Announcements continued to speculate that Edison's 'happy combination of electricity and photography' hailed the coming of the telephonoscope, not the cinema.[67] A correspondent for *Harper's* reported, 'Mr. Edison holds that with the kinetograph and the telephone combined he has reproduced, visually and audibly, a theatrical performance many miles away from the scene of the actual prediction'.[68] *Life* magazine joked that, with the Kinetograph, the telephone user would finally be able to look upon the face of 'that sweet-voiced operator'.[69]

Popular magazines and national newspapers, such as *Illustrated American*, *Harper's*, and *Leisure Hour*, copied the tone established by the *Sun*. Much

65 'The Kinetograph,' *New York Sun*, 28 May 1891.

66 'Punch and the Kinetoscope,' *Illustrated American*, 21 June 1891, 224.

67 'Edison's Kientograph,' *Engineering*, 5 June 1891; 'Edison's Latest Invention: The Kinetograph,' *Leisure Hour*, August 1891, 711-712.

68 'Electricity at the Fair,' *Harper's Weekly*, 16 July 1892.

69 'We May Be Disappointed,' *Life*, 2 July 1891, 412. See also 'Local Gossip,' *Trewman's Exeter Flying Post or Plymouth and Cornish Advertiser* (UK), 13 May 1893.

of the coverage reinforced the *Sun*'s confidence that Edison's Kinetograph would live up to all the hype and then some. For example, *Cassell's Family Magazine* referred to death as a 'disability' that could be overcome by having one's face and voice recorded for perpetuity on the kinetograph.[70] Unlike the cynical reception of the 'Far-Sight Machine', the Kinetograph received overwhelming praise.

Still, several reports attempted to balance the otherwise uncritical fanfare. For example, *American Engineer* commented: 'Thomas A. Edison is not half so garrulous as the crudely manufactured highly imaginative interviews published from time to time would lead the public to believe.'[71] The *Chicago Journal* adopted a tone similar to the *Boston Journal*'s response to the 'Far-Sight Machine'. Speculating on the kind of world the Kinetograph would create, the journalist strikes out at the invalid in the same manner earlier writers had criticized the 'man who goes out between the acts'. Anxious of the social change the Kinetograph would bring forth, the journalist goes on to say: 'As for the performers themselves, how would they get among before a vacant unlit house speaking or singing towards the darkness and hearing no applause. It is to be feared such performance would be lifeless.'[72] The *Journal*'s recognition of the social effects of the Kinetograph fell on deaf ears. At least in mainstream newspapers and magazines, the overwhelmingly progressive rhetoric of Edison's supporters surpassed the criticisms lodged against his new invention.

Discussion of the Kinetograph took on a much different tone in science journals like *Engineering* and *American Engineer*. Letters to the editor of *English Mechanic* saw it for what it really was—a derivative attempt to re-brand the work that had already been done by lesser-known electricians. Most correspondents wrote in attempting to parse the so-called originality of Edison's idea. Letters printed throughout June and July chime in on the many other inventors -- including Muybridge, Marey, and Anschutz -- who had already presented inventions for 'instantaneous photography'.[73] Still

70 'The Kinetograph,' *Cassell's Family Magazine*, August 1891, 575-576.

71 'Reflection on Western Journalism,' *American Engineer* (attributed to *Electrical Enterprise*), 6 June 1891, 223.

72 'What Edison Claims,' *Chicago Journal*, 13 May 1891.

73 Jebus Bickle, 'Edison's Kinetograph,' letter to the editor, *English Mechanic*, 19 June 1891, 359-360; Eldridge, letter to the editor, *English Mechanic*, 3 July 1891, 406; Jabez Ogle, 'The Kinetograph,' letter to the editor, *English Mechanic*, 24 July 1891, 479; W. A. Rudge, 'The Kinetograph and Other Optical Apparatus,' letter to the editor, *English Mechanic*, 3 July 1891, 406; T. R., 'Seeing by Electricity,' letter to the editor, *English Mechanic*, 31 July 1891: 504-505.

others rejected the invention outright, confused about Edison's claims to its electrical nature.[74]

> It seems to me that Mr. Edison must suffer muchly from the 'greatness' which has been thrust upon him [....] What in the name of common sense are 'powerful' reflectors, and how can you convey anything from them – whether 'powerful' or not – by means of wires which are misnamed 'electric'?[75]

These letters point out the fact that the kinetograph has nothing whatsoever to do with either telegraphy or telephony. Several correspondents referenced the contemporary work of Shelford Bidwell, noting the discrepancy between Edison's invention for photographic reproduction, called 'instantaneous photography', and Bidwell's work on image transmission, called 'telegraphic photography'. The progress over the course of the discussion shows how these electricians were able to disentangle the notion of 'instantaneous photography' from that of 'telegraphic photography', more commonly known as 'seeing by electricity.' But these insights was lost on the wider public.

Over the course of the next few years, journalists, writers, and scientists continued to hone in on the Kinetoscope, negotiating expectations of 'seeing by electricity' with the realities of photographic and phonographic recording. Only later, as demonstrations were underway in 1893, did it become clear what the machine could actually do: record living scenes on photographic filmstrips.[76] The ambiguities existed in the gap between the public's expectations, based on existing technologies like the phonograph and the telephone, and perceived utilities, like home theatre.

Even while the photographic nature of the Kinetograph moved into the foreground, the idea of the cinema never entirely lost its association with electricity. As the language of the cinema continued to develop, emphasizing the realism, natural likeness, fidelity, and mechanical precision, its electric

74 E. August, 'Seeing by Electricity,' letter to the editor, *English Mechanic*, 3 July 1891, 409; Nun Dor, 'The Kinetograph,' letter to the editor, *English Mechanic*, 10 July 1891, 431; Eldridge, letter to the editor (reply to Nun Dor), *English Mechanic*, 24 July 1891, 479; 'Seeing by Electricity,' *English Mechanic*, 31 July 1891, 504-505; E. August, 'Seeing by Electricity,' letter to the editor (Reply to T. R.), *English Mechanic*, 28 August 1891.

75 Nun Dor, 'The Kinetograph,' letter to the editor, *English Mechanic*, 10 July 1891, 431.

76 'First Public Exhibition of Edison's Kinetograph,' *Scientific American*, 20 May 1893; 'Edison's Kinetograph,' *The Photographic News*, 16 June 1893.

liveness continued to play a vital role.[77] The cinema's electric character brought a sense of liveness, spectacle, and vibrancy to a recording medium otherwise understood as dead and lifeless.[78]

Edison made his ultimate claim to the invention of cinema in his successful case against the American Mutoscope company in 1898.[79] This case also demonstrates the development of a unique identity for the cinema as a combination of electricity and photography. By presenting the argument that Edison had 'invented' the 'art of living pictures', distinct from the Kinetographic camera, lawyer Richard Dyer steered the argument away from the formal mechanics and towards its ambiguous intellectual property. Edison's case relied on proving that the inventor's novel contribution was one of *practice*, not of *mechanical construction*. The repetition of keywords and key phrases creates the impression of an *active presence* as opposed to a *passive recording*: 'the art of *living* pictures', 'fidelity of *life*', 'illusion of *movement*', and so forth.[80] The testimony documents the impact that cultural expectations, not to mention language itself, can have on the genesis of new technology. Even if the cinema failed to establish an actual connection, as Edison had promised in its electrical association with the telegraph and telephone, it at least would be able to present *a realistic illusion* of the presence of a person or 'living scene'.

77 Electric liveness could be thought of as a kind of electrical sublime in line with David Nye's examination of late nineteenth-century technology and culture. See David Nye, *American Technological Sublime* (Cambridge, MA: MIT Press, 1996); David Nye, *Electrifying America: Social Meanings of a New Technology, 1880-1940* (Cambridge, MA: MIT Press, 1992).

78 'What Edison Claims,' *Chicago Journal*, 13 May 1891; Mr. Edison's Kinetograph,' *Illustrated American*, 20 June 1891, 194.

79 Dickson, who had been largely responsible for developing the Kinetograph and Kinetoscope, left Edison's employ in 1895 to form the Mutoscope Company. See: Paul Spehr, 'Filmmaking at the American Mutoscope and Biograph Company 1900—1906,' *The Quarterly Journal of the Library of Congress* 37, no. 3/4 (Summer/Fall 1980), 413; Charles Musser, 'Before the Rapid Firing kientograph: Edison film production, representation and exploitation in the 1890s,' in *Edison Motion Pictures, An Annotated Filmography 1890-1900* (Washington, D. C.: SI Press, 1997); Hendricks, *The Kinetoscope*; Hendricks, *The Origins of Motion Pictures*; Spehr, *The Man Who Made Movies*; Musser, *The Emergence of Cinema*, 303-313; Musser, *Before the Nickelodeon*; *Studios before the System*. On the outcome of the case, see 'Continued Legal Battles,' in *A Guide to Motion Picture Catalogs by American Producers and Distributors* (Frederick, MD: University Publications of America, 1985). This source gives the dates and the rulings; Mutoscope appealed and overturned the decision, granted March 1902.

80 Edison v. Mutoscope, 31, 41, 66, 74, 81, 200, 227, 256.

THE KINETOSCOPE ON THE BOWERY, NEW YORK CITY—VIEWING A "KNOCK-OUT."

Figure 20. Antonia Dickson, 'Wonders of the Kinetoscope,' *Frank Leslie's Illustrated Magazine*, March 1895, 256, 257.

Conclusion: 'Electric' Cinema

The 'Far-Sight Machine', that electric periscope that could twist every which way, clashed with traditions of privacy, security, and social presence. The 'Far-Sight Machine's' association with electric liveness worked to its detriment. Paradoxically, for the Kinetograph, that 'happy combination of electricity and photography' worked to its advantage. Promoting the Kinetograph as a device for the home implied a dark, private viewing space.[81] Imbuing that image with electricity enlivened the idea that the reproduction of a scene could come to seem natural.

Comparing the 'Far-Sight Machine' as depicted in Figure 16 with the Kinetoscope in Figure 20 illustrates this conclusion. Both show users interacting with a peephole viewer with the image seen composed in a round frame within the frame. In the article, Dickson describes the Kinetoscope as a '"seeing" machine':

> Its functions are to give us the representation of life, not as the painting, the photograph or the statue represents it, frozen into a single attitude,

81 Kinetograph announcements in 1891 present the invention as a device for the home.

> but exhibiting all the wealth of movement and expression which makes up the sum of out restless existence [....] The combined effect is life, with all its eloquent and insistent appeals to the senses of man.[82]

From this perspective, the cinema certainly has more in common with the notion of 'far-sight' than it does with photography. For Dickson, as with moving image discourse in general, the appearance of lifelike movement overrode the distinction between live and recorded images.

Transitions such as this may be more apparent to an historian than to daily newspaper reporters and their reading audience. The curious transformation that took place, as the cinema grew up in the shadow of the tradition of 'seeing by electricity', became more evident as the years rolled on. Just as the best critiques of American exceptionalism came from the British, so did the best retrospectives. In 1896, magician Nevil Maskelyne reflected on the short history of the animated picture in an interview for the British penny magazine *To-Day*.[83] While Maskelyne was less than optimistic about the commercial future of animated pictures, he operated London's Egyptian Hall, according to film historian John Barnes, only the second hall in England ever to be used as a cinema.[84] Maskelyne acknowledged the similarities between the 'far-sight machine' and the Kinetoscope, expressing disappointment with the Kinetograph and hope that the 'far-sight machine' would soon make an appearance.

> 'You seem almost disappointed with the [Kinetoscope] machine, Mr. Maskelyne?'
> 'Well, I am, in more ways than one. It is not new; the principle is not new. When we heard that Edison was bringing out a "far-sight" machine, we all thought it was going to be something very much more elaborate than the Kinetoscope. Of course, in one sense, the Kinetoscope might be called a "far-sight" machine, but it is not at all what we had expected it would be.'
> 'And what were you hoping for?'

82 Antonia Dickson, 'Wonders of the Kinetoscope,' *Frank Leslie's Illustrated Magazine*, March 1895, 245.
83 Louise Kane, "*To-Day* Has Never Been 'Highbrow'': Middlebrow, Modernism, and the Many Faces of To-Day,' in *Transitions in Middlebrow Writing, 1880-1930*, eds. Kate Macdonald and Christopher Singer (New York: Palgrave MacMillan, 2015), 58.
84 John Barnes, *The Beginnings of Cinema in England 1894-1901* (London: University of Exeter Press, 1997), 117; Richard Steven Cohn, 'Who Put the Magic in Movies?' *SMPTE Motion Imaging Journal* (February-March 2006), 86; Antonia Lant, 'The curse of the Pharaoh, or how cinema contracted Egyptomania,' *October* 59 (1992): 87-112.

> 'A real "far-sighting" machine--an instrument that will do for the eye what the telephone has done for the ear. It is to be a machine by the use of which a man will be enabled to see his friends although they may be many hundreds of miles away from him. This is what the scientific world is waiting for and so we were a trifle disappointed with the Kinetoscope.'
> 'Do you think this "far-sighting" instrument will ever be invented?'
> 'Certainly I do!'[85]

While the American press was caught up in the enthusiasm for the Kinetograph and the Kinetoscope, a different tone set in across the Atlantic. Perhaps because of the lag in the time it took for information to travel, or because of their closer proximity to the cinema scene in Paris, the British journals approached the subject of 'seeing by electricity', the Kinetoscope, and the mania over Edison's latest inventions with acute scepticism. Memory of technological failure persisted in England, as reviewers, weary of Edison's rhetoric and exaggerated claims, held strongly to their critical lens. When W. K. L. and Antonia Dickson published a monograph in praise of Edison's life and work, the British reviewers were the first to note the inconsistencies in their stories.[86] In England, Edison turned into the 'Electric Barnum', whose notoriety served to obscure the fact of his dishonesty. The reviewer attacked the Dicksons' attempt to call Edison 'the greatest genius of this or any other age', calling the authors out for their exaggerated claims, many of which had no basis in fact.

85 'Mr. Maskelyne and the Animated Photograph Craze,' *To-Day: A Weekly Magazine-Journal*, 16 May 1896, 39.

86 W. K. L. Dickson and Antonia Dickson, *The Life and Inventions of Thomas Alva Edison* (TY Crowell, 1894); 'The Electric Barnum,' review of *The Life and Times of Thomas Alva Edison*, by W. K. L. and Antonia Dickson, *The Saturday Review of Politics, Literature, Science and Art*, 1 December 1894, 601-602; 'The Life and Work of Edison,' review of *The Life and Inventions of Thomas Alva Edison*, by W. K. L. and Antonia Dickson, *The Dial*, 16 November 1894, 289-291; Review of *The Life and Inventions of Thomas Alva Edison*, by W. K. L. and Antonia Dickson, *The Electrical Engineer*, 28 November 1894, 443; Review of *The Life and Inventions of Thomas Alva Edison*, by W. K. L. and Antonia Dickson, *The Electrical World*, 23 February 1895, 245; Review of *The Life and Inventions of Thomas Alva Edison*, by W. K. L. and Antonia Dickson, *Popular Science*, February 1895, 559; Henry Tyrell, 'Edison,' *Frank Leslie's Popular Monthly* (March 1895), 258-271; Review of *The Life and Inventions of Thomas Alva Edison*, by W. K. L. and Antonia Dickson, *The Engineering and Mining Journal*, 20 July 1895, 51; Frank Mundell, *The Story of Edison and the Wonders of Electricity* (Jarrod & Sons, 1898); Review of *The Story of Edison and the Wonders of Electricity*, by Frank Mundell, *The Spectator*, 20 November 1897, 713; 'The Electrical Barnum Again,' *The Saturday Review*, 20 November 1897, 560; 'Our Booking Office,' review of *The Story of Edison*, by Frank Mundell, *Punch*, 27 November 1897, 250.

> Thus the reader is directly invited to suppose that Edison invented the duplex method of telegraphy, which was in reality in use before he was born. That he improved the quadruplex telegraph and perfected it is not disputed; though he did not originate even this. And as to the octuplex [sic] system, there is no such thing in existence yet.[87]

The reviewer put the Dicksons' claims in historical context by recalling that, in 1878, Edison's announcement about the invention of the electric light had contributed to a nationwide financial crisis, a controversy of which Americans seemed ignorant. British commentators overall presented more of an interest and a capacity to dethrone Edison and to combat his advocates' false claims. Tales such as 'Electric Barnum' is one of the best that technological folklore has to offer, relaying a critical position to combat the insular American attitude towards progress, the negative aspects of technological progress, and the continuities of technological invention as social practice.

But the invention of the cinema, as it emerged in America in the late nineteenth century, involved a confluence of discursive forces that contributed to its cultural identity. The popular rejection of the 'Far-Sight Machine' speaks to cultural values of visibility, privacy, and chivalry. Attacks lodged against stock characters such as the 'man who goes out between the acts' and the cheating husband indicates a distaste for surreptitious actions and shirking of social responsibilities. At the same time, the 'Far-Sight Machine's' 'fierce white light' threatened to reveal intimate knowledge. Together, these criticisms note firm distinctions between public and private spaces, along with the appropriate behaviours attributed to each setting.

Understanding early cinema culture through the lens of *media in transition* reveals the ways in which it built upon established customs, practices, and institutions. Edison's introduction of the Kinetograph rode both implicitly and explicitly on expectations established by 'seeing by electricity'. In the process of its emergence, the construction of cinema contributed to a reformulation of expectations about 'seeing by electricity'. The 'art of living pictures' that evolved in early cinema culture, characteristic of a way of thinking about moving images as a particular medium, fulfilled some but not all of the promises that Edison had initially made. The cinema would make it possible to stop time, but not to annihilate space.

87 'The Electric Barnum,' 601.

Bibliography

Aberdeen Weekly Journal. 'Edison's Coming Invention.' June 15, 1889.

The American (Waterbury, CT). 'Untitled' (Edison's Far-Sight Machine). May 22, 1889.

American Engineer. 'Engineering and Manufacturing Notes.' May 7, 1890.

American Engineer. 'Reflection on Western Journalism.' June 6, 1891, 223.

American Journal of Photography. 'Edison's Latest' (reprinted from the *Philadelphia Daily News*). June 1891, 338.

American Settler. 'Chat.' July 12, 1890.

Atchison Daily Champion (Atchison, KS). 'Edison's Machines.' Aug 10, 1889.

The Atlanta Constitution (Georgia). 'People Here and There.' February 14, 1890.

August, E. 'Seeing by Electricity,' letter to the editor (Reply to T. R.), *English Mechanic,* August 28, 1891.

Baltimore Herald. 'Untitled (Edison's Far-Sight Machine).' June 10, 1889.

Barnes, John. *The Beginnings of the Cinema in England 1894-1901.* London: University of Exeter Press, 1997.

Barnum, P.T. 'Notes and Comments: What the Fair Should Be.' *North American Review.* March 1890, 400-401.

Bickle, Jebus. 'Edison's Kinetograph,' letter to the editor, *English Mechanic,* June 19, 1891, 359-360.

The Big Sandy News (Louisa, KY). 'The Kinetograph: With it you can see a man a thousand miles away—Edison's Latest.' April 28, 1890.

Birmingham Daily Post. 'Gleanings' (One of the most peculiar of Mr. Edison's recent inventions). June 15, 1889.

Boston Journal. 'Edison's Last.' May 13, 1889.

Brief for Complainant, Thomas A Edison v. The American Mutoscope Company and Benjamin F. Keith, 1898 (TAED QM001195).

Brooklyn Journal. 'Edison's Talk.' October 7, 1889.

Carpenter, Frank. 'Edison in his Den.' *The Pittsburg Dispatch.* November 5, 1891.

Cassell's Family Magazine. 'The Kinetograph.' August 1891, 575-6.

Chambers' Journal. 'The Month – Science and Arts.' July 25, 1891, 477.

The Chicago Evening Post. 'Edison's in Chicago: The Wizard of Menlo Park Stopping at the Auditorium, Tells of his Latest Invention.' May 12, 1891.

Chicago Journal. 'What Edison Claims.' May 13, 1891.

Chicago Tribune. 'Edison and the Big Fair.' May 14, 1891.

Cohn, Richard Steven. 'Who Put the Magic in Movies?' *SMPTE Motion Imaging Journal* (February/March 2006): 80-94.

The Commercial (Portland, IN) 'Untitled' (Edison's Far-Sight Machine). September 5, 1889.

Commercial Gazette (Cincinatti, OH). 'Untitled' (Edison's Far-Sight Machine). July 18, 1889.

Conot, Robert *A Streak of Luck: The Life and Times of Thomas A. Edison*. Boston, MA: Da Capo Press, 1979.

The Courier and London & Middlesex Counties Gazette. 'A Far-Sight Machine.' June 22, 1889.

The Democrat (Hutchinson, KS). 'Untitled' (Edison's Far-Sight Machine). August 17, 1889.

The Dial. 'The Life and Work of Edison,' review of *The Life and Inventions of Thomas Alva Edison*, by WKL and Antonia Dickson. November 16, 1894.

Dickson, Antonia. 'Wonders of the Kinetoscope.' *Frank Leslie's Illustrated Magazine*. March 1895.

Dickson, W.K.L. and Antonia Dickson. *History of the Kinetograph, Kinetoscope, and Kineto-phonograph*. New York: Albert Bunn, 1895; New York, Museum of Modern Art 2000.

Dickson, W.K.L. and Antonia Dickson. *The life and inventions of Thomas Alva Edison*. New York: TY Crowell, 1894.

Dor, Nun. 'The Kinetograph,' letter to the editor, *English Mechanic*, July 10, 1891, 431.

Eldridge, E. letter to the editor (reply to Nun Dor), *English Mechanic*, July 24, 1891, 479.

Eldridge, E., letter to the editor, *English Mechanic*, July 3, 1891, 406.

Electrical Engineer. 'The Edison Kinetograph.' June 24, 1891, 708.

Electrical Engineer. 'Edison's Photophonokinetograph,' May 20, 1891, 584.

Electrical Engineer. 'Notes: Electric Seeing.' May 30, 1890.

Electrical Engineer. Review of *The Life and Inventions of Thomas Alva Edison*, by W.K.L. and Antonia Dickson, November 28, 1894, 443.

Engineering. 'Edison's Kinetograph.' June 5, 1891, 678.

Electrical Review. 'A Far-Sight Machine.' May 25, 1889.

Electrical Review. 'A Far-sighted Individual.' August 2, 1889, 126.

Electrical Review. 'How Electricity Will Help the Editor of the Future.' Feb 2, 1889, 4.

Electrical World. 'The Kinetograph.' June 13, 1891, 431.

Electrical World. 'The Kineto-phonograph.' June 16, 1894.

Electrical World. Review of *The Life and Inventions of Thomas Alva Edison*, by W.K.L. and Antonia Dickson. February 23, 1895.

Engineering and Mining Journal. Review of *The Life and Inventions of Thomas Alva Edison*, by W.K.L. and Antonia Dickson, July 20, 1895.

Engineering and Mining Journal. 'Edison, The Inventor of the Phonograph.' Aug 1878, 131.

English Mechanic. 'Edison's Kinetograph' (reprinted in the *Times* London). June 5, 1891, 310.

English Mechanic. 'Edison's Photophonokinetograph.' May 20, 1891.
The Evening World (New York). 'The Kinetograph.' May 28, 1891.
Far-Sight Machine, Menlo Park Scrapbook, 1889 (TAED SM035).
Fun. 'Goaheadison's Real Latest.' July 17, 1889.
Fun. 'The Long and Short of It.' June 19, 1878.
Fun. 'Professor Goaheadison's Latest.' July 3, 1889.
Gaudreault, André, ed. *American Cinema, 1890-1909: Themes and Variations.* New Brunswick, NJ: Rutgers University Press, 2009.
Gaudreault, Andre. *Film and Attraction: From Kinematography to Cinema.* Chicago: University of Illinois Press, 2011.
Gunning, Tom. 'An Aesthetic of Astonishment: Early Film and the (In)credulous Spectator.' *Art and Text* 34 (Spring 1989): 31-45.
Gunning, Tom. 'The Cinema of Attraction.' *Wide Angle* 3, no. 4 (1986).
Gunning, Tom. 'Doing for the Eye What the Phonograph Does for the Ear.' In *The Sounds of Early Cinema*, edited by Richard Abel and Rick Altman. Bloomington: Indiana University Press, 2001.
Gunning, Tom. "Primitive' Cinema: A Frame-up? Or the Trick's on Us.' *Cinema Journal* 28, no. 2 (1989): 3-12.
Harper's Weekly. 'Electricity at the Fair.' July 16, 1892.
Hendricks, Gordon. *The Edison Motion Picture Myth.* Berkeley: University of California Press, 1961.
Hendricks, Gordon. *The Kinetoscope: America's First Commercially Successful Motion Picture Exhibitor.* New York: Beginnings of the American Film, 1966.
Herbert, Stephen. *A History of Early Television.* New York: Taylor & Francis, 2004.
Herbert, Stephen. 'Professor Goaheadison's latest.' *Early Popular Visual Culture* 9, no. 1 (2011): 75-81.
Hidalgo, Santiago. 'Early American Film Publications: Film Consciousness, Self Consciousness.' In *A Companion to Early Cinema*, edited by André Gaudreault, Nicolas Dulac, and Santiago Hidalgo, 202-219. New York: Wiley, 2012.
Hunt, Verity. 'Electric Leisure: Late Nineteenth-Century Dreams of Remote Viewing by Telectroscope.' *Journal of Literature and Science* 7, no. 1 (2014): 55-76.
Illustrated American. 'Between the Acts.' October 4, 1890, 3.
Illustrated American. 'Mr. Edison's Kinetograph.' June 20, 1891, 194.
Illustrated American. 'Punch and the Kinetoscope.' June 20, 1891, 224.
Illustrated London News (American edition, NY). 'Untitled,' August 19, 1889.
Indianapolis Herald (Indiana). 'Light Wedded to Sound.' May 31, 1891.
Iron. 'The Past Year. A General Retrospective.' January 3, 1890.
Iron. 'What Next?' June 14, 1889.
Jacobson, Brian R. 'The Black Maria: Film Studio, Film Technology.' *History and Technology* 27, no. 2 (2011): 233-241.

Jacobson, Brian R. *Studios Before the System: Architecture, Technology, and Early Cinema.* New York: Columbia University Press, 2015.

Jenkins, Reese Valmer, Charles Musser, and Thomas E. Jeffrey. *A Guide to motion picture catalogs by American producers and distributors, 1894-1908: a microfilm edition.* University Publications of America, 1985.

Journal (Lincoln, NE). 'Untitled' (Edison's Far-Sight Machine). May 26, 1889.

Journal (Pottsville, PA). 'Untitled' (Edison's Far-Sight Machine). June 19, 1889.

Judy. 'A Run on 'Far-Sight Machines." June 26, 1889.

Kane, Louise. "*To-Day* Has Never Been 'Highbrow': Middlebrow, Modernism, and the Many Faces of To-Day.' In *Transitions in Middlebrow Writing, 1880-1930*, edited by Kate Macdonald and Christopher Singer. New York: Palgrave MacMillan, 2015.

Kluitenberg, Eric, ed. *The Book of Imaginary Media: Excavating the Dream of the Ultimate Communication Medium.* Rotterdam, The Netherlands: NAi Publishers, 2007.

Kluitenberg, Eric. 'On the Archaeology of Imaginary Media.' In *Media Archaeology: Approaches, Applications, and Implications*, edited by Erkki Huhtamo and Jussi Parikka, 48-69. Berkeley: University of California Press, 2011.

Lant, Antonia. 'The Curse of the Pharaoh, or How Cinema Contracted Egyptomania.' *October* 59 (1992): 87-112.

Lastra, James. *Sound Technology and the American Cinema: Perception, Representation, Modernity.* New York: Columbia University Press, 2000.

Lathrop, George Parsons. 'Edison's Kinetograph.' *Harpers Weekly.* June 13, 1891, 446-447.

Lathrop, George Parsons. 'If Edison Can Find the Way.' *The Youth's Companion.* January 10, 1895.

The Leader (Wilkes Barre, PA). 'Untitled' (Edison's Far-Sight Machine). August 9, 1889.

Leisure Hour. 'Edison's Latest Invention: The Kinetograph.' August 1891, 711-712.

Levant Herald. 'Mr. Edison and the Electric Millennium.' September 1, 1889.

Life. 'We May Be Disappointed.' July 2, 1891, 412.

Litigation Records Series, Thomas A. Edison v. American Mutoscope Company, 1898-1900 (TAED QM001).

Lloyd's Weekly Newspaper (UK). 'Edison's Latest Invention.' June 16, 1889.

Lockwood, Thos. D. 'Observations.' *The Electrical Engineer* (US). October 1889, 423.

Manufacturer and Builder. 'Edison's Kinetograph.' June 1891, 138.

Marvin, Carolyn. 'The Electrical Imagination: Predicting the Future of Communications in Britain and the United States in the Late Nineteenth Century.' PhD diss., University of Illinois at Urbana-Champaign, 1979.

Milwaukee Daily Journal (Milwaukee, WI). 'Personal and Impersonal' (If Edison had his Far-sight machine in order). May 24, 1889.

Milwaukee Daily Journal (Milwaukee, WI). 'Wizard of Menlo Park.' April 26, 1890.

Moore, Paul. 'The Social Biograph: Newspapers as Archives of the Regional Mass Market for Movies.' In *Explorations in New Cinema History: Approaches and Case Studies,* edited by Richard Maltby, Daniel Biltereyst, and Philippe Meers, 263-279. Hoboken, NJ: Wiley-Blackwell, 2011.

Morning Advertiser (London). 'Untitled' (With all its 'hem's and 'ha's). February 4, 1890.

Morning Oregonian. 'Electric Marvels: An Apparatus for the eye as the phonograph is to the ear.' November 16, 1890.

Mundell, Frank. *The Story of Edison and the Wonders of Electricity.* London: Jarrod & Sons, 1897.

Musser, Charles *Before the Nickelodeon: Edwin S. Porter and the Edison Manufacturing Company.* Berkeley, CA: University of California Press, 1991.

Musser, Charles. *The Emergence of Cinema: The American Screen to 1907.* Berkeley, CA: University of California Press, 1994.

Natale, Simone and Gabriele Balbi, 'Media and the Imaginary in History: The Role of the Fantastic in Different Stages of Media Change.' *Media History* 20, no. 2 (2014): 203-218.

New York Graphic. 'Untitled' (Edison's Far-Sight Machine.' June 26, 1889.

New York Herald. 'Mr Edison at Home Unspoiled by Glory.' October 7, 1889.

New York Herald. 'To Catch a Speaker's Gestures.' February 2, 1890.

New York Sun. 'Edison's Conjury.' May 13, 1891.

New York Sun. 'The Kinetograph: Edison's latest and most surprising device.' May 28, 1891.

New York Sun. 'To See by Electricity: An Apparatus for the Eye as the Phonograph is for the Ear' (attributed to *Iron*). September 12, 1890.

New York Times. 'An Authentic Life of Edison,' review of *The Life and Inventions of Thomas Alva Edison*, by W.K.L. and Antonia Dickson. November 11, 1894.

New York Times. 'A Move in the Right Direction – Mr. Edison's Latest.' May 29, 1891.

New York Times. 'A World's Fair in 1892.' June 25, 1889.

News and Citizen (Morrisville, VT). 'Untitled' (One of the curiosities of Edison). February 13, 1890.

Nye, David. *Electrifying America: Social Meanings of a New Technology, 1880-1940.* Cambridge, MA: MIT Press, 1992.

Nye, David. *American Technological Sublime.* Cambridge, MA: MIT Press, 1996.

Ogle, Jabez, 'The Kinetograph,' letter to the editor, *English Mechanic*, July 24, 1891, 479.

Pall Mall Gazette (London). 'Fourth Edition' (Edison's Latest—A Far-Sight Machine). June 15, 1889.

Pall Mall Gazette (London). 'Mr. Edison's Latest Invention.' May 29, 1891.

Pall Mall Gazette (London). 'Occasional Notes.' May 29, 1891.
Pall Mall Gazette (London). 'Untitled (Edison has added a new horror to existence).' February 3, 1890.
Perrysberg Journal. 'Untitled' (Edison at the Electric Light Convention). February 15, 1890.
The Post (Liverpool, England). 'A Far-Sight Machine.' June 17, 1889.
Phonogram. 'The Kinetograph.' October 1892, 217-219.
The Photographic News. 'Edison's Kinetograph.' June 16, 1893.
Popular Science. Review of *The Life and Inventions of Thomas Alva Edison,* by WKL and Antonia Dickson. Feb 1895.
Punch. 'Open House (To be dated after the next invention).' August 10, 1889.
The Record (West Chester, PA). 'Untitled' (Edison's Far-Sight Machine). July 18, 1889.
The Republican (Meridan, CT). 'Untitled.' (Edison's Far-Sight Machine). May 14, 1889.
Rudge, W.A. 'The Kinetograph and Other Optical Apparatus,' letter to the editor, *English Mechanic,* July 3, 1891, 406.
Sacramento Daily Record. 'Edison's Latest.' February 3, 1890.
Sacramento Daily Record. 'Latest from the Wizard.' February 1, 1890.
Sandage, Scott. *Born Losers: A History of Failure in America.* Cambridge, MA: Harvard University Press, 2005.
The Saturday Review. 'The Electric Barnum,' review of *The Life and Times of Thomas Alva Edison,* by W.K.L. and Antonia Dickson, December 1, 1894, 601-2.
The Saturday Review. 'The Electrical Barnum Again.' November 20, 1897, 560.
Scientific American. 'A Far-Sight Machine.' June 1, 1889.
Scientific American. 'First Public Exhibition of Edison's Kinetograph.' May 20, 1893.
Scientific American. 'Edison's Kinetograph.' June 20, 1891, 393.
Scott, George B., 'Edison's Latest Marvel: The Kinetograph,' letter to the editor. *English Mechanic,* June 19, 1891, 358-359.
Segal, Howard. *Technological Utopianism in American Culture.* Syracuse, NY: Syracuse University Press, 1985.
Segal, Howard. *Future Imperfect: The Mixed Blessings of Technology in America.* Amherst, MA: University of Massachusetts Press, 1994.
The Spectator. Review of *The Story of Edison and the Wonders of Electricity,* by Frank Mundell, November 20, 1897.
Spehr, Paul. 'Filmmaking at the American Mutoscope and Biograph Company 1900—1906.' *The Quarterly Journal of the Library of Congress* 37, no. 3-4 (Summer/Fall 1980), 413-421.
Spehr, Paul. *The Man Who Made Movies: W.K.L. Dickson.* New Barnet, UK: John Libbey Publishing, 2008.
St. James Gazette. 'Was Mr. Edison in Earnest?' June 12, 1889.
The Statesman (Yonkers, NY). 'A Far-Sight Machine.' June 11, 1889.

Steele, Penn. 'Anticipations: How Some of Mr. Wells' Speculative Predecessors have Fared.' *The Era Magazine,* Jan-Jun 1902, 460-468.

Strauven, Wanda, ed. *The Cinema of Attractions Reloaded.* Amsterdam: Amsterdam University Press, 2006.

T.R. 'Seeing by Electricity,' letter to the editor, *English Mechanic,* July 31, 1891: 504-5.

The Time (Bethlehem, PA). 'Untitled' (Edison's Far-Sight Machine). June 25, 1889.

The Times (London). 'Mr. Edison and the Chicago Exhibit.' May 14, 1891.

The Times (London). 'Mr. Edison's Latest Invention.' May 29, 1891.

The Times (Portsmouth, NH). 'Untitled' (Edison's Far-Sight Machine). July 29, 1889;

To-Day: A Weekly Magazine-Journal. 'Mr. Maskelyne and the Animated Photograph Craze.' May 16, 1896.

Townsend, Horace. 'Edison: His Work and his Workshop.' *Cosmopolitan* (April 1889), 598-607: 602.

Tribine (Racine, OH). 'Untitled' (Edison's Far-Sight Machine). July 24, 1889.

Tyrell, Henry. 'Edison,' *Frank Leslie's Popular Monthly.* March 1895.

Uricchio, William. 'Storage, Simultaneity, and the Media Technologies of Modernity.' In *Allegories of Communication: Intermedial Concerns from Cinema to the Digital,* edited by John Fullerton and Jan Olsson, 123-138. Bloomington: Indiana University Press, 2004.

Uricchio, William. 'Television, Film and the Struggle for Media Identity.' *Film History* 10, no. 2 (1998): 118-127.

Uricchio, William and Roberta Pearson. 'Coming to Terms with New York City's Moving Picture Operators, 1906–1913.' *The Moving Image: The Journal of the Association of Moving Image Archivists* 2, no. 2 (2002): 73-93.

Wahpeton Times (Wahpeton, ND). 'A Dead Madman's Words.' February 13, 1890.

Washington Post. 'The American Wizard: More Wonderful Things that Edison is to Bring Forth.' September 1, 1889.

Western Electrician (Chicago). 'Edison's Visit to Chicago.' May 23, 1891, 295.

Western Electrician (Chicago). 'A Scotch Estimate of Edison.' January 5, 1895.

Western Mail (UK). 'Edison's Far-Sight Machine.' June 15, 1889.

Wheeling Daily Intelligencer (Wheeling, WV). 'Untitled' (Edison's promised Kinetograph). May 25, 1891.

The World (Omaha, NE). 'Untitled' (Edison's Far-Sight Machine). May 24, 1889.

Yenowine's News (Milwaukee, WI). 'Gossip on the Quiet.' June 23, 1889.

Zielinski, Siegfried. *Audiovisions: Cinema and Television as Entr'actes in History.* Amsterdam University Press, 1999.

3. Human-Seeing Machines

From Annihilating Space to Mediated Vision

Abstract

Chapter Three investigates the emergence of systems thinking in the historical development of television around the turn of the 20th century. This chapter elucidates the continuities and discontinuities linking the Victorian conception of 'seeing by electricity' and the Machine-Age construction of electronic screens. The scientific developments that facilitated electronic technology and the sociopolitical philosophy of efficiency contributed to a new conception of television. The rhetoric of the annihilation of space that had propelled nineteenth-century progress was displaced by a belief that human beings should adapt to these new, artificial environments. Comparing and contrasting extension theory and systems thinking shows how this new philosophy of technology contributed to a new way of thinking about 'distant electric vision'.

Keywords: Annihilation of space; progress; philosophy of technology; systems thinking; efficiency

Predictions of a future in which people would see by electricity did not die out despite fundamental technical barriers. These fantasies persisted into the first decade of the 20th century. While seeing by electricity continued to be a persistent concept during television's speculative era, it also contributed to expectations for television when it re-emerged in American popular culture in the early 20th century. Television as we know it, the electronic transmission of moving images, came about in the 20th century. A. A. Campbell Swinton coined the term 'distant electric vision' in a 1908 letter published in *Nature*, which introduced electronics and applied physics

---

Roberts, I., *Visions of Electric Media: Television in the Victorian and Machine Ages.*
Taylor & Francis Group, 2019
DOI 10.5117/9789462986596_CH03

into the practices of seeing by electricity.[1] Swinton's letter indicates a major shift in the scientific community's approach to engineering and designing television. It encouraged engineers to adopt electronics in their designs. As a result, the forms of television changed to incorporate human physiology and the role of the human observer in these new systems.

Examining the similarities and differences between seeing by electricity in nineteenth-century visual culture and television in the early 20th century reveals the ways in which technology was thought to mediate communication and visual perception. Similar arguments have been made about the transformation of vision in visual culture. Jonathan Crary and Stephen Kern, for example, have examined the intellectual history, science, and literature demonstrative of a shift in ways of seeing between the nineteenth and 20th centuries.[2] Studies in the visual culture of science, such as Nicolas Rasmussen's and Oliver Gaycken's histories of the microscope, also challenge the apparently self-evident definition of media and practices of image-making.[3] Martin Willis also examined Victorian literature and science to demonstrate a similar kind of shift.[4] In film history scholarship, William Uricchio and Tom Gunning have also made claims about the role of modernity in shaping a new kind of vision and thus contributing to the culture and practice of the cinema.[5]

Seeing by electricity encompasses a history of literature and culture without actual material invention. For this reason, historians most often relegate this period to the 'pre-history' or speculative era populated with

1 A. A. Campbell Swinton, 'Distant Electric Vision,' letter to the editor, *Nature* 78, no. 2016 (18 June 1908): 151.

2 Jonathan Crary, *Techniques of the Observer: On Vision and Modernity in the Nineteenth Century* (Cambridge, MA: MIT Press, 1992); Stephen Kern, *The Culture of Time and Space, 1880-1918* (Cambridge, MA: Harvard University Press, 1983).

3 Nicolas Rasmussen, *Picture Control: The Electron Microscope and the Transformation of Biology in America, 1940-1960* (Stanford, CA: Stanford University Press, 1999); Oliver Gaycken, '"The Swarming of Life": Moving Images, Education, and Views through the Microscope,' *Science in Context* 24, no 3 (2011): 361-380.

4 Martin Willis, *Vision, Science and Literature, 1870-1920: Ocular Horizons* (New York: Routledge, 2015).

5 William Uricchio, 'Phantasia and Technè at the Fin-de-siècle,' *Intermédialités: Histoire et théorie des arts, des lettres et des techniquesIntermediality:/History and Theory of the Arts, Literature and Technologies* 6 (2005): 27-42; Tom Gunning, 'The World as Object Lesson: Cinema Audiences, Visual Culture and the St. Louis World's Fair, 1904,' *Film History* 6, no. 4 (1994): 422-444; Tom Gunning, 'Re-newing Old Technologies: Astonishment, Second Nature, and the Uncanny in Technology from the Previous Turn-of-the-century,' in *Rethinking Media Change: The Aesthetics of Transition,* eds. Henry Jenkins and David Thornburn (Cambridge, MA: MIT Press, 2003), 39-60.

'ego-documents' and science fiction.[6] Even though the culture of seeing by electricity produced no functional media artefacts, its visual culture and popular science offer glimpses into the formation of expectations for both cinema and television. Including the cultural and imaginative dimensions of technology with the technical and scientific offers a broader, more inclusive, and more robust view of the processes through which cultures and technologies co-emerge.

By focussing closely on the historical period 1878-1911, this chapter examines the visual culture and discourse of television. The visual culture of seeing by electricity bears resemblance to the 'television' that emerged in the first decade of the 20th century. By examining the similarities and differences between these two types of television, I will show how a way of seeing thought to be unmediated transformed into a new construction of vision dominated by realistic illusions and screen-mediated communication.

The cultural construction of mediated vision in both cases closely aligns with two distinct philosophies of technology. The nineteenth-century culture of seeing by electricity promoted a vision of technology as facilitating the user's ability to extend the body through space. The popular rhetoric was summed up in the familiar phrase 'the annihilation of space', and supports a philosophy of technology associated with machines as extensions of the body. Seeing by electricity visualized the extension of the eye through space, able to see over the physical horizon and access distant points instantaneously. With the emergence of large technical systems and electronic practices, the mode of engineering changed to one of systems.[7] Whereas nineteenth-century designs emphasized devices analogous to parts of the body, 20th-century systems incorporated the process of human vision into the technical methods for seeing by electricity. Along with the burgeoning sciences of psychology and physiology and the progressive efficiency movement, 'distant electric vision' reworked television as a kind of seeing and the observer as a human seeing-machine. Television works because scientists engineered a new way of seeing that relied as much on human visual and cognitive perception as on the mechanics of electricity and light.

6 Andreas Fickers, 'Television,' in *The Handbook of Communication History*, eds. Peter Simonson et al. (New York: Routledge, 2013), 239; Russell W. Burns, 'Part One: The Era of Speculation 1877 to c. 1922,' in *Television: An International History of the Formative Years* (London: IET, 1998), 3-140; George Shiers, *Early Television: A Bibliographic Guide to 1940* (New York: Routledge, 1997), x.

7 Thomas Hughes, 'The Evolution of Large Technological Systems,' in *The Social Construction of Technological systems: New Directions in the Sociology and History of Technology,* eds. Wiebe E. Bijker, Thomas Hughes, and Trevor Pinch (Cambridge, MA: MIT Press, 1987): 51-82.

This chapter examines the transition from the nineteenth-century visual culture of seeing by electricity to the emergence of systems thinking in the early 20th century. After identifying the philosophy of technology associated with the nineteenth-century rhetoric of space-annihilation, this chapter examines the early 20th-century developments in engineering. The theory that technology extends innate human capacities evolved from a basic one-to-one relationship between eye and device to a more sophisticated understanding of visual and technological systems. Electronic engineering contributed to new directions in television development. Thinking about television as a large technical system enabled electronic engineering to reconfigure the shape and meaning of this new technology. A renewed faith in engineering made way for a vision of the human and the machine working together as parts of an efficient system.

Seeing by Electricity, Annihilating Space

Of the dozens of engineers who worked on the problem of seeing by electricity, only a handful were recognized in both technical and popular communities: Constantin Senlecq, George Carey, Shelford Bidwell, and Jan Szczepanik.[8] Senlecq and Szczepanik called their inventions 'telectroscopes', while Carey used the term 'selenium camera', and Bidwell coined the phrase 'telegraphic photography'. These engineers stand out because their names show up most frequently in the literature in both major daily newspapers like the *London Times* and the *New York Sun* as well as scientific periodicals such as *Nature* and *Scientific American*. These four inventors also published visual depictions of their schematics, an extra element that gave their work a more recognizable component. The seeing-by-electricity craze reached its height in the 1880s. All sorts of frauds and tricksters crawled out of the woodwork with claims that they had constructed a working mechanism. Several were revealed to be hoaxes and those that were not surely were just empty promises and grandstanding. Scientists were taken more seriously when their claims were accompanied by a visual design, schematic, or other demonstrative component.

8 On Senlecq, see 'Telectroscope,' *Le Mondes*, 16 January 1879, 90; 'The Telectroscope,' *Times* (London), 27 January 1879 (reprinted in *Nature, The Electrician, Scientific American,* and *English Mechanic*). On Carey, see 'Seeing by Electricity,' *Scientific American*, 5 June 1880, 355. See also Shelford Bidwell, 'Telegraphic Photography,' *Journal for the Society of Telegraph Engineers and Electricians* 10 (September 1881): 357; Shelford Bidwell, 'Telephotography,' *Nature*, 10 February 1881, 344-346; 'Herr Szczepanik's Telectroscope,' *New York Tribune*, 23 March 1898.

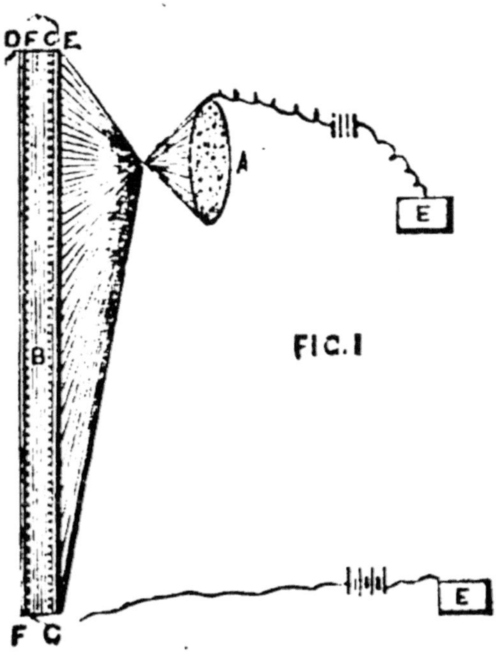

Figure 21. Constantin Senlecq, 'The Telectroscope' *The Electrician,* 5 February 1881, 141-142.

Designs offered by Senlecq, Carey, and Szczepanik most accurately characterize the general concept of seeing by electricity as it was known in nineteenth-century scientific and popular culture. French scientist Constantin Senlecq was the first to present a 'telectroscope' to a popular audience in 1879, just weeks after *Punch* published the spread attacking Edison for his talk of electric light (Figure 21).[9] Modelled after the camera obscura and powered by electricity, Senlecq's device took advantage of state-of-the-art knowledge about the electrical conductivity of the mineral selenium. News of Senlecq's telectroscope spread fast and wide, making appearances in *London Times, New York Sun,* and *Scientific American.* His name resurfaces two years later when, in 1881, he published a revised and updated design.[10] Senlecq was nothing if not persistent.

9 'Telectroscope,' *Le Mondes,* 16 January 1879, 90; 'The Telectroscope,' *Times* (London), 27 January 1879 (reprinted in *Nature, The Electrician, Scientific American,* and *English Mechanic*); 'The Telectroscope,' *Telegraphic Journal,* 15 February 1879; 'The Telectroscope Used for Photographing Over Telegraph Wires,' *The Manufacturer and Builder,* April 1879, 86; 'Editor's Scientific Record: Miscellaneous,' *Harper's New Monthly Magazine,* May 1879, 947.

10 Constantin Senlecq, 'The Telectroscope,' *The Electrician,* 5 February 1881, 141-142; 'The Telectroscope,' *English Mechanic,* 11 February 1881, 534-535; 'The Telectroscope,' *Scientific*

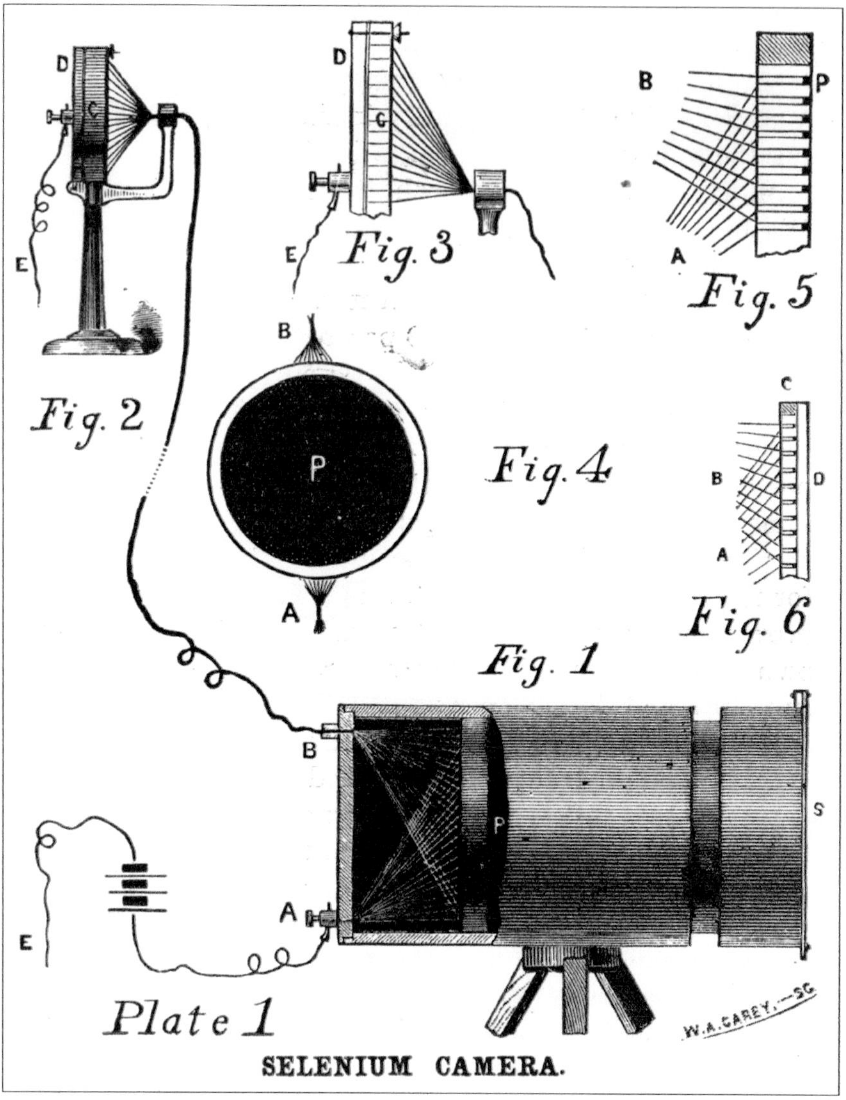

Figure 22. George Carey's Selenium Camera, 'Seeing by Electricity,' *Scientific American*, 5 June 1880, 355.

American amateur inventor George Carey also published sketches for a 'selenium camera' in *Scientific American* in 1880 (Figure 22).[11] Despite the name, his designs resemble Senlecq's closely. Selenium cells turn the light

American Supplement, 9 April 9, 1881, 4382.

11 'Seeing by Electricity,' *Scientific American*, 5 June 1880, 355.

into electrical voltage, which is then transmitted along 100 wires to individual cells assembled in a mosaic in the distant screen. Both sketches show the screen in profile in order to provide the optimal view of the electrical circuitry. On one end, the camera obscura captures the light reflecting off of a scene. Picturing the screen from its side seems counter-intuitive to anyone familiar with television as a visual medium. But, in 1879, there were no moving images. Instead, these depictions drew attention to the electrical wiring. Emphasizing the similarities between the mosaic of cells and the retina, inventors and journalists appealed to readers' knowledge about the eye to describe how television would work.

If anyone had any luck at all, it was Shelford Bidwell. He recognized early on that the mechanisms involved would only be able to transmit still pictures, which is why he patented his process as 'telegraphic photography' (Figure 23). The dream of television persisted despite the physical odds. And people continued to draw pictures of what a mechanical eye would look like. But the history of how scientists figured out ways to transmit still pictures (facsimile) diverges from the history of moving-image technology when these distinct technologies began producing positive results in the early 20th-century.[12]

After a decade of failed experiments, hoaxes, and hype, a sense of disillusionment set in among the scientific community about the possibility of seeing by electricity. By the 1890s, Bidwell became recognized as the authority on the science of seeing by electricity. A discussion in *English Mechanic*, for example, referred to his work as the most promising accomplishment in the history of attempts to reproduce images at a distance.[13] One correspondent made a distinction between Bidwell's transmission of still pictures and the possibility of transmitting moving images. But, because

12 Shelford Bidwell, 'Telegraphic Photography,' *Journal for the society of telegraph engineers and electricians* 10 (September 1881): 357; Shelford Bidwell, 'Telephotography,' *Nature*, 10 February 1881, 344-346. On the history of facsimile transmission, see Jonathan Coopersmith, *Faxed: The Rise and Fall of the Fax Machine* (Baltimore, MD: Johns Hopkins University, 2015). On physiological optics, see Doron Galili, 'Chapter II: Unexpected Organs: Television, Modern Vision, Modernist Aspirations,' in 'Seeing by Electricity: The Emergence of Television and the Modern Mediascape, 1878-1939,' PhD diss., University of Chicago, 2011. On the relationship between light and vision, see Chris Otter, *The Victorian Eye: A Political History of Light and Vision in Britain, 1800-1910* (Chicago, IL: University of Chicago Press, 2008); Sydney Perkowitz, *Empire of Light: A History of Discovery in Science and Art* (New York: Henry Holt, 1996); Martin Jay, *Downcast Eyes: The Denigration of Vision in Twentieth-century French Thought* (Berkeley, CA: University of California Press, 1993); Arthur Zajonc, *Catching the Light: The Entwined History of Light and Mind* (Cambridge, UK: Oxford University Press, 1995).

13 T. R., 'Seeing by electricity,' letter to the editor, *English Mechanic*, 31 July 1891, 503-504; 'Seeing by electricity,' letter to the editor, *English Mechanic*, 26 June 1891, 387-388; E. August, 'Seeing by Electricity,' letter to the editor, *English Mechanic*, 3 July 1891, 409.

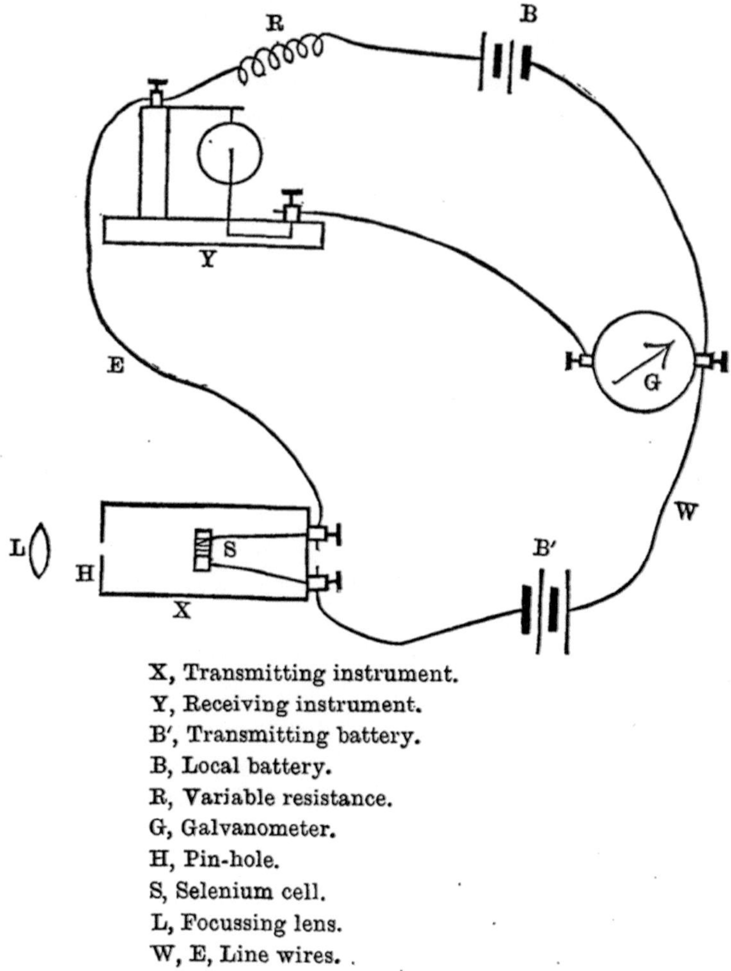

Figure 23. Shelford Bidwell, 'Telegraphic Photography,' *Journal for the Society of Telegraph Engineers and Electricians* 10, September 1881, 357.

of the technical nature of Bidwell's work, he did not receive much notoriety outside of the specialized British community of technicians devoted to the practical science of telegraphy. Any discussion of 'seeing by electricity' tended to attract characters more prone to flights of fancy.

Nearly two decades after the telectroscope had first been introduced, Polish inventor Jan Szczepanik made his claim to fame with the new telectroscope, a favourite in the European as well as American press leading up to the 1900 Paris Exposition Universelle. Szczepanik's grandiose claims and his intriguing biography made him an attractive rags-to-riches tale, a story that the press flocked to when taking advantage of the hype over the upcoming Exposition (Figure 24). As a result, Szczepanik drew far more attention in the popular press than in technical publications. In contrast to Senlecq and Carey, Szczepanik's designs took a different form. Szczepanik's telectroscope utilized mirrors, and descriptions of the device appealed to the reproduction of images, not to selenium cells. Comparing these two generations of telectroscopes shows how the invention of the cinema had already started to have an effect on the field of electrical engineering. Other inventors of this time were known to have replaced the familiar selenium with mirrors and projection screens, perhaps to resemble the new invention of cinema more closely or else simply to downplay selenium as yesterday's news.

The vigour with which the press promoted Szczepanik's telectroscope resembles the attention generated by Edison's inventions.[14] One journalist remarked how Szczepanik would 'out Edison Edison'.[15] Along with the enthusiasm for the young Polish inventor came a similar rhetoric. While Edison's journalists recycled themes of wizardry and go-aheadism, Szczepanik became associated with the power to extend one's grasp across the vastness of space by means of technology.

Linking Szczepanik with Edison's recognizable 'Far-Sight machine', a Boston journalist hailed his telectroscope as the 'latest step toward space-annihilation'.[16] The phrase 'the annihilation of space', though already popularized decades earlier by telegraphic journalists, came back with full force in stories about Szczepanik. As Stephen Kern explains it:

14 International press covered Szczepanik for two years. See 'Gooseberries in March,' *Electrical Engineer*, 4 March 1898, 257; 'Next, Please!' *Electrical Engineer*, 11 March 1898, 304-305; 'The Far-Seer,' *Boston Daily Advertiser*, 21 March 1898; 'Herr Szczepanik's Telectroscope,' *New York Tribune*, 23 March 1898; 'Telectroscopy,' *Electrical Engineer*, 25 March 1898, 354; Dr. Johannes Horowitz, 'That New Telectroscope,' *New York Times*, 3 April 1898; 'The Fernseher Again,' *Electrical Engineer*, 15 April 1898: 449; 'More Szczepanik,' *Electrical Engineer*, 22 April 1898: 483; 'The Latest Triumph of Electricity,' *Illustrated London News*, 23 April 1898; 'Genius' Triumph over Great Obstacles,' *The Milwaukee Journal*, 23 April 1898; 'Science and Discovery,' *Weekly Rocky Mountain News*, 5 May 1898; 'Inventor of the Telectroscope,' *Weekly Rocky Mountain News*, 12 May 1898; 'Personal and Impersonal,' *Milwaukee Daily Journal*, 24 May 1889; 'A Great Invention,' *The Bristol Mercury and Daily Post*, 2 June 1898; 'More Szczepanik,' *Electrical Engineer*, 3 June 1898: 675; 'The German Press and Szczepanik,' *Electrical Engineer*, 1 July 1898, 3; 'The Telelectroscope and its Inventor,' *American Monthly Review of Reviews*, 18 July 1898, 93-94; 'More Szczepanik,' *Electrical Engineer*, 29 July 1898, 129; Mark Twain, 'The Austrian Edison Keeping School Again,' *The Century*, August 1898, 630-631; Cleveland Moffett, 'Seeing by Electricity,' *Pearson's Magazine*, October 1899, 490; Mark Twain, From the 'London Times' of 1904, Century, November 1898, 100-105; 'Paris Exhibition,' *Glasgow Herald*, 12 April 1900.

15 'Inventor of the Telectroscope,' *Weekly Rocky Mountain News*, 12 May 1898.

16 'The Far-Seer,' *Boston Daily Advertiser*, 21 March 1898.

Figure 24. Jan Szczepanik and his marvelous invention. 'The Latest Triumph of Electricity,' *Illustrated London News*, 23 April 1898, 593.

> The 'annihilation of distance' was not a science-fiction fantasy or some theoretical leap of physicists; it was the actual experience of the masses who quickly became accustomed to an instrument that enabled them to raise money, sell wheat, make speeches, signal storms, prevent log jams, report fires, buy groceries, or just communicate across ever increasing distances.[17]

Conventionally used to hail the extraordinary advances in science, technology, and industry, 'the annihilation of space' associated the new sense of domination over physical and natural limitations made possible by railroads, telegraphic networks, and the telephone.

Historian of technology Leo Marx identified the trope in his important 1964 book *The Machine in the Garden*, which linked the American tradition of the pastoral with the ironic fascination with machines. Marx wrote:

> No stock phrase in the entire lexicon of progress appears more often than the "annihilation of space and time," borrowed from one of [Alexander] Pope's relatively obscure poems [....] The extravagance of this statement apparently is felt to match the sublimity of technological progress.[18]

17 Stephen Kern, *The Culture of Time and Space*, 214-215.

18 Leo Marx, *The Machine in the Garden: Technology and the Pastoral Ideal in America* (Cambridge, UK: Oxford University Press, 1964), 194. See also Howard P. Segal, *Technological Utopianism in American Culture* (Syracuse, NY: Syracuse University Press, 1985); David Nye, *Electrifying America: Social Meanings of a New Technology, 1880-1940* (Cambridge, MA: MIT Press, 1992); David Nye, *American Technological Sublime* (Cambridge, MA: MIT Press, 1996); Kenneth Goldberg, ed. *The Robot in the Garden: Telerobotics and Telepistemology in the Age of the Internet* (Cambridge, MA: MIT Press, 2001).

In this passage, Marx identifies several tropes in the rhetoric of American technological progress, including associations between the machine, nature, and history relevant to nineteenth-century American literature. Marx explains how the machine and nature seemed to fuse together into a 'technological sublime'. His analysis helps explain the exaggerated rhetoric that fuelled the culture of seeing by electricity. The many claims about the invention of seeing by electricity implicitly connected the power of new technology with a sense that humankind had gained mastery over nature. The telectroscope would make it possible to extend a person's vision beyond the physical limitations established by nature.

Along with the rhetoric of technological progress associating Szczepanik with the annihilation of space, his popularity also brought the German philosophy of technology to the awareness of the English-speaking world. The papers made Szczepanik out to be a sort of follower of the work of R. E. Liesegang, and, by extension, fashioned Liesegang into a kind of guru. Several articles noted how Szczepanik's interest in electrical engineering grew from the inspiration he found in the writings of Liesegang.[19] Liesegang's philosophy of technology, which promoted the power of technology to make humans stronger, was even more extreme and progressive than that of his American counterparts. The introduction reads like a manifesto: 'When the first automaton, that is better constructed than man, is brought to life, the purpose of the world will have been achieved: Man will be God.'[20] In a sense, Liesegang's treatise fuses the philosophy of technology with the popular science of television.

By the time Szczepanik rose to fame, Liesegang had already published several books, including *Die Organologie* (*Organology*) (1892) and *Beiträge zum Problem des elektrischen Fernsehen* (*Contributions to the problem of electric television*) (1891; 1899).[21] *Organology* laid the foundation for his philosophy of technology, an 'attempt to eliminate the dualism between organic and inorganic'.[22] In *Contributions*, Liesegang outlined not only the technical requirements for the functioning of television, but also a philosophical way of thinking about the meaning of technology. Along with noted German

19 'Der Elektrische Ferneher, besuch bei Hern Ein Jan Szczepanik,' *Neue Weiner Tageblatt* (Vienna, Austria), 17 March 1898; 'Das Telelectroscop' *Lanterna Magica* 14.54 (Leipsig), May 1898.

20 R. E. Liesegang, *Beiträge zum Problem des elektrichen Fernsehen* (Dusseldorf, 1891; 1899); Siegfried Zielinski, *Audiovisions: Cinema and Television as Entr'actes in History* (Amsterdam: Amsterdam University Press, 1999), 133.

21 On Liesegang, see Stefan Andriopolous, 'Psychic Television,' *Critical Inquiry* 31, no. 3 (2005): 632; Zielinski, *Audiovisions*, 32.

22 *Die Organology*, quoted in Ernst A. Hauser, 'Raphael Eduard Liesegang 1869-1947 (Necrology),' *Journal of Chemical Education* 26, no. 5 (May 1949): 274.

philosopher Ernst Kapp, Liesegang promoted a philosophy of technology rooted in the idea that machines extend humankind's natural abilities.[23] Liesegang's 'organology' drew on Kapp's theory of 'organ projection', following Kapp's 'extended argument that all technical artifacts are projections of human organs, in that "humans unconsciously transfer form, function and the normal proportions of their body to the works of their hands."'[24] Kapp had coined the new term the 'philosophy of technology', and the idea that machines were extensions of the human body in both concept and design became the basis for *extension theory*. In the 20th century, prominent American communications scholar Marshall McLuhan popularized the media as 'extensions of man'.[25] These theories encourage a way of thinking about technology as *prosthetics*. This approach posits that people model tools after embodied faculties, as the hammer extends the arm. In this way, technology extends, supplements, or replaces parts of the body.

While relatively obscure in American media history, German literature scholar Stefan Andriopoulos noted Kapp and Liesegang's important contributions to the German history of television in his study on early television philosophy:

> Liesegang opens his *Contributions on the Problem of Electrical Television* with a reference to Kapp's *Outlines of a Philosophy of Technology*, according to which 'almost all tools, machines, etc. are unconscious copies that imitate parts of the human being' (*P*, p. iii). Liesegang, for whom the Morse telegraph corresponded to the human sense of touch and the telephone to the ear, thus understood his 'instrument for the telegraphing of lens-produced images' as 'imitating the sense of sight' (*P*, pp. 1, iv).[26]

23 On Kapp, see Pasi Väliaho, *Mapping the Moving Image: Gesture, Thought and Cinema circa 1900* (Amsterdam: Amsterdam University Press, 2010), 80-82; Carl Mitcham, *Thinking Through Technology: The Path Between Engineering and Philosophy* (Chicago, IL: University of Chicago Press, 1994), 21-23; Philip Brey, 'Technology as Extension of Human Faculties,' in *Metaphysics, Epistemology and Technology*, ed. Carl Mitcham (London: JAI, 2000), 7-8.

24 Brey, 3 (quoting Kapp 1877, p. v-vi, Brey's translation); Ernst Kapp, *Grundlinien einer Philosophie der Technik* (Braunschweig, Germany: Druck und Verlag von George Westermann, 1877). On the German 'Machine-Age culture' and the philosophy of technology, see Heidi Voskuhl, 'Engineering Philosophy: Theories of Technology, German Idealism, and Social Order in High-Industrial Germany,' *Technology and Culture* 57, no. 3 (2016).

25 Marshall McLuhan, *Understanding Media: The Extensions of Man* (Cambridge, MA: MIT Press, 1964).

26 Andriopoulos, 632.

According to Andriopoulos, a direct line can be traced from the emergence of the philosophy of technology in Germany in the late nineteenth century and the technical development of television occurring at the same time. If Liesegang drew on Kapp, and Szczepanik found inspiration in Liesegang, their philosophy of technology found its way into the English-speaking world by association. Andriopoulos encourages an inclusive view of the history of television culture:

> The slow accumulation of technical and physical knowledge, beginning around 1890, accelerating in the 1920s, and enabling the first wireless transmissions of moving pictures in the last years of that decade did not take place in a vacuum that could be separated from its contingent cultural contexts.[27]

While Andriopoulos focusses on the connections between occultism and television history, a similar point should be made about the way historical modes of thinking about technology represented in the philosophy of technology provide models for both the cultural reception of new technology and the technological development of new inventions. In this way, the late nineteenth-century popularity of extension theory links to the rhetoric of technological progress found in the 'annihilation of space' as well as the shape and meaning of 'seeing by electricity' in American popular culture.

The analogy supporting the philosophy of extension theory came across in both the verbal descriptions and the visual depictions of seeing by electricity. The particular approach taken by these engineers emphasizes the electrical function of the device. Technical descriptions detail the mechanism by which selenium converts light into electricity. Special care is taken to describe the process by which the devices would transmit light in the form of electricity. One strategy resurfaces in technical explanations linking the technical design with the eye's retina. Engineers would liken their diagram to the arrangement of rods and cones in the retina, for example. Irish inventor Denis Redmond described his electric telescope functioning like a human eye in his 1878 letter published in *English Mechanic*:

> By using a number of circuits, each containing selenium and platinum arranged at each end, just as the rods and cones are in the retina, the

27 Andriopoulos, 622.

> selenium end being exposed in a camera, I have succeeded in transmitting built-up images of very simple luminous objects.[28]

Describing their devices for seeing by electricity with reference to the faculty of vision proved a common way of explaining not only how the technology was meant to work, but also how it would change the way we see. Drawing the connection between the electric seeing device and the human eye carried with it an implicit assumption about the relationship between man and machine. An electric telescope was more than a tool: it would help move the human observer towards the goal of annihilating space.

Similarly, a letter published in the *London Times* that year corroborated the association between eye and electric telescope. Middleton of St. John's College outlined a lecture recently given before the Cambridge Philosophical society:

> [I] pointed out a striking analogy between the camera of the instrument and that of the human eye ; the thermoelectric elements of the instrument and the rods and cones of the eye ; the conducting system of insulated wires emanating from the plate of the instrument and the optic nerve (or bundle of conducting fibres of the eye) – supposing that as the electric currents in the instruments effected a registration on the sensitive paper, so in the eye the nerve currents to the optic nerve probably leave some brain trace on the mind.[29]

Drawing on physiological metaphors to explain the process of seeing by electricity had the double advantage of humanizing a technical craft and bringing a recognizable function to the proposed technology. Journalists and inventors alike persisted in explaining the meaning of seeing by electricity by reference to the way the devices resembled and, to an extent, were modelled after the human eye.

Represented in both verbal descriptions and visual diagrams, these designs emphasize the eye-camera analogy, a method quite distinct from the emphasis on visible images and screens that emerged following the popularity of the cinema. The visual culture of seeing by electricity emphasizes process over picture for the very reason that these devices only existed on paper. Reproducing moving images was simply not possible

28 'An Electric Telescope,' letter to the editor, *English Mechanic*, 7 February 1879.

29 H. Middleton, 'Seeing by telegraph,' *Times* (London), 24 April 1880. See also 'Seeing by Electricity,' *The Electrician*, 7 March 1890, 448-450.

using nineteenth-century methods for manufacturing selenium, referred to as 'sluggish'.[30] These designs were radically impractical in both cost and operation. Were someone to have built a working prototype, it would have cost an estimated 1.25 million pounds, and even then the synchronization and speed of transmission would have been insufficient to process a true moving picture.[31] The telectroscope was never actually built, and, as a result, no one ever had the pleasure of looking through an electric telescope to see the world beyond the horizon.

Engineers tended to depict their designs for seeing by electricity with emphasis on the stand-alone devices. This strategy supports the method of descriptions used, which focussed on the processes of electrical transmission and mechanical synchronization. In the schematics accompanying the technical descriptions and in artist visualizations that appeared in the illustrated news, telectroscopes, electric telescopes, and selenium cameras also bear a physical resemblance to the design of the human eye. In contrast to the fantastic literature and satires depicting the culture of telephonoscope, which emphasize magnificent screens and magic mirrors, the technical literature represents seeing by electricity as a mechanical endeavour built on the principles of electrical engineering. Little attention is paid to the screens or to the images that were expected to appear in them. Instead, descriptions and schematics focus on the placement of electrical wiring and mechanical construction of the equipment.

Two particular designs stand out that illustrate the analogy between the human eye and the electric telescope. The first, a 'selenium eye' invented by Werner von Siemens, offers a literal translation of the eye into a scientific device meant to simulate vision. The other, a mosaic of selenium cells, represents a common approach to the design of the receiver: a nineteenth-century version of a television screen. Schematics in patent applications and diagrams in technical periodicals alike portray the devices, whether a selenium camera or a mosaic screen, like an 'artificial retina' or an 'electrical eye', with selenium (light-responsive) cells assembled like the rods and cones

30 Marcus Martin, 'Television,' *Electricity*, 21 October 1921, 577; Shelford Bidwell, 'Telegraphic photography and electric vision,' letter to the editor, *Nature* 78, no. 2014 (1908): 105-106; A. A. Campbell Swinton, 'Distant electric vision,' letter to the editor, *Nature* 78, no. 2016 (1908): 151.

31 For cost estimates, see Shelford Bidwell, 'Telegraphic Photography and Electric Vision.' For practical details on speed and synchronism, see 'Seeing by Electricity,' *The Electrician*, 7 March 1890, 448-450; Shelford Bidwell, 'Telephotography,' *Nature* 23, no. 589 (February 1881): 344-346; Shelford Bidwell, 'Telegraphic photography,' *Journal of the Society of Telegraph Engineers and Electricians* 10, no. 38 (September 1881): 354-360.

in the retina.[32] These writers concentrated less on the role of the actual observer or user of the device and more on the resemblance between the technology and the body. Sometimes the devices resembled eyes in their design as well as in their discursive explanation, as in the case of Werner von Siemens's literal approach to the artificial eye (Figure 25). *Scientific American* described the illustration of Siemens's 'electric eye': 'the whole is comparable to an eye, in which the screens represent the lids, and the selenium plate the retina'.[33] A quote from the inventor follows, giving the analogy more concrete form:

> 'Here,' says Dr. Siemens, 'is an artificial eye, sensible to light and to differences in colour, which gives signs of fatigue when it is submitted to the prolonged action of light, which regains its strength after resting with closed lids,' and which, by an electro-magnet attachment, may be made to close itself, as does the human eye involuntarily, on the occurrence of a vivid flash.

Siemens's electric eye provided an early model for thinking about the analogy between the eye and the seeing-machine in a very literal way. While it may not have been a device for seeing like the other telectroscope schemes, it illustrates the design philosophy connecting the limitations of human physiology with the powers of technology to defy nature.

While the cathode ray tube provides the recognizable model for the television screen, its nineteenth-century counterpart was designed using the eye as a model. Engineers described the construction of mosaic screens of selenium cells that resembled the arrangement of rods and cones in the retina. Among the many proposals, Fritz Lux's 1902 patent application provides an exemplary model (Figure 26). While Lux wrote of the inspiration he drew from nature, he also notes the limits to the metaphor.

> To construct an apparatus that works in practice, it is best to take as the role model nature, that produces such wonderful and perfect faculties. Suppose taking nature's eye as a model for the construction of a television. Even so, the model does not inform on the actual process of seeing, but one can assume that it determines that the image projected on the retina is

32 Werner Von Siemens, 'Action of Light on Selenium,' *Nature* 13, no. 334 (March 1876); 'Siemens' Sensitive Electric Eye,' *Scientific American*, 8 December 1876: 374.

33 'Siemens' Sensitive Electric Eye,' *Scientific American*, 8 December 1876: 374.

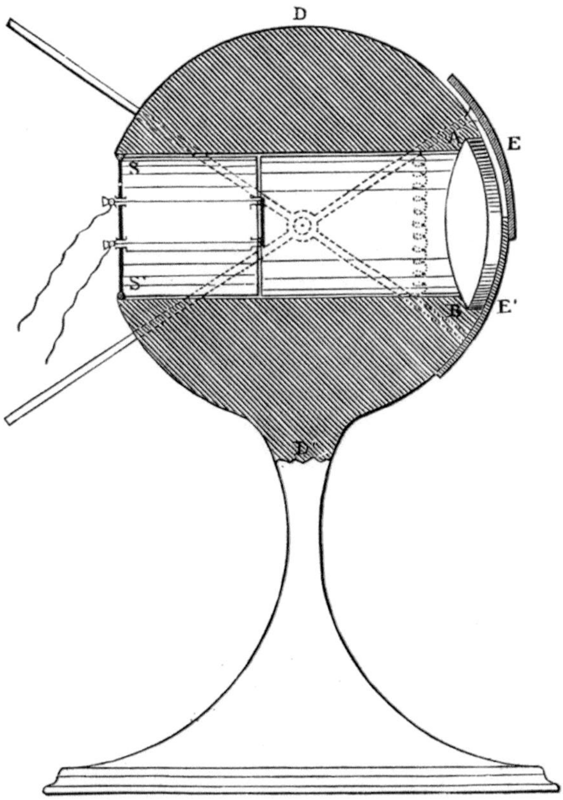

Figure 25. 'Siemens' Sensitive Electric Eye', *Scientific American*, 8 December 1876, 374.

> transmitted instantaneously to the brain. And so it is also the imperative with television to transmit the image instantaneously.'[34]

Siegfried Zielinski refers to Lux's mosaic as an 'archaic pixel structure'.[35] For a contemporary reader, the resemblance between the electrical and organic mosaics would have been clearly apparent. The analogy maintained a strong presence in both the design of technical artefacts and the engineer's methods of description.

Reference to the body and the eye in particular continued to support explanations of television after the turn of the century. These analogies persisted into the 1920s, and, to some degree, have never entirely left the discourse. During the height of the seeing-by-electricity craze in the 1880s,

34 Fritz Lux, *Der Elektrische Fernseher* (Ludwigshafen, 1903), 7.
35 Zielinski, *Audiovisions*, 71.

Figure 26. Fritz Lux, *Der Elektrische Fernseher.* Ludwigshafen, 1902.

less attention was paid to the screen and to the image. That changed after the invention of the cinema, as the emergence of moving-image discourse made the persistence of vision and the role of the viewer increasingly important in the functioning of the visual illusion. Instead of the 1:1 correlation between eye and camera, the analogy persisted as a common explanation to which was added a host of scientific explanations and user experiences. As the popular science of television developed into the first decades of the 20th century, the discussion continued to draw on the body-machine analogy, while ocular physiology and the process of human vision became more and more important in explaining how and why television worked.

Extension theory has become an enduring aspect of the philosophy of technology because of the simplistic way it connects the human and the machine. The approach supports the enduring metaphor that technological design resembles organically evolving organisms. One could not hope for a more commonsensical explanation of the meaning of technology than extension theory's underlying principle: the hammer is designed as an extension of the arm. The appeal of extension theory is seen in the popularity of Marshall McLuhan's work. His version of extension theory promotes a vision of the body as incomplete without a technological supplement, a prosthetic.

To interpret extension theory merely as a theory of prosthetics strips the philosophy of its underlying basis in the duality of man and machine. Doron Galili, for example, in his study of extension theory in early 20th-century film and television, notes how the theory of prosthetics works as a model for television as easily in its nineteenth-century speculative era as in its early 20th-century technological development.[36] Insofar as the theory of the prosthetic functions as an analogy of vision that links the eye to the mechanical camera, it draws from the simple body-machine metaphor. Film scholar Pasi Valaiho, on the other hand, provides a denser interpretation of extension theory that recognizes its roots in the duality of man and machine.[37] Valaiho's reading of Kapp's theory of organ projection comes to resemble the theory of the cyborg, which recognizes the human as a hybrid entity. A theory of prosthetics based on the body-machine metaphor treats the human as if the body was an incomplete organism without some kind of technological supplement. The extension theory of Kapp, on the other hand, developed out of the belief that man and machine were more alike than different.

Identifying correlations in the cultural history of television with the development of a popular science and a philosophy of television provides a more robust strategy. Restricting extension theory to a culturally specific late nineteenth-century construction of power, progress, and space annihilation allows for a distinction to be drawn between the late nineteenth-century philosophy of technology and the enduring metaphors linking the human body to mechanical designs. Extension theory arose in full force in the cultural climate of the late nineteenth century, associated with the power of technology to annihilate space. Kapp's philosophy of technology, for example, grew up from a foundational understanding of the body and the

36 Galili, 109.
37 Valaiho, 80-82.

machine as dual mechanisms. Without the underlying belief in human-machine hybridity, 20th-century extension theory came to rely too heavily on a simplistic body-machine metaphor. Beyond the metaphor linking the human body to the design of technical artefacts, however, the theory falls short of explaining the complexities of television.

Extension theory has its limits, however, as a means for explaining the cultural aspects of technological change. Cultural history provides an alternative way of understanding technological change, distinct from a focus on the history of technology. From this view, technological developments can be understood as reflecting cultural attitudes about the limits of the human.

Extension theory reflects the late nineteenth-century conception of technological change. Since extension theory became popular in the late nineteenth-century, the technologies that grew up during this time could also be understood as reflecting these same attitudes towards the limits of the human body's natural capacities – to hear and see at a distance, to record pictures automatically.

From this perspective, the technological ability to hear and see at a distance, and even to capture pictures automatically either by photographic or cinematic means, could be understood as signifying the cultural desire to compensate for a lack of such capacities in the human body itself. For example, the telephone compensates for the body's incapacity to extend indefinitely across space; while the telephone allows its users to hear over vast distances, it also reveals their innate limitation fixed as we are in space and time. Extension theory also suggests that there is something artificial or unnatural about technology. If people make tools to compensate for a lack, then technology provides the means to make people more than human. Its proponents have been accused of technological determinism, a single-minded view that relegates 'progress' to the force of technological change.[38] Extension theory in general and the notion of technology as prosthetics more specifically suggests that tools lend power to the humans who wield them, eventually allowing for the few to rule over the many. This view encapsulates an outdated Victorian conception of technology. It suggests an imbalanced relationship between technology and the self that is powered by the ambition to rule the world, or otherwise to *annihilate space.* When science imposes the sense of deficiencies in vision, in the form of faulty eyes

38 Van Loon, 28-29. See also Paul Levinson, *Digital McLuhan: A Guide to the Information Millennium* (New York: Routledge, 1998).

and slow senses, technology comes to save the day. It considers the human, at worst, flawed and incomplete, and, at best, a hybrid organic machine.[39]

As historian of technology Carl Mitcham has shown, Kapp's philosophy of technology grew out of a particular cultural and historical moment. As such, it conveys a German historical-materialist approach to technology: 'Along with Marx, Kapp was a left wing Hegelian[...]. Kapp's adaptation of Hegelian dialectic called for the 'colonization' and transformation of this environment, both internally and externally.[40] To adopt a conception of technology that views technics as a form of life entails a consideration of the ways such theories reflect on cultural and historical constructions of human identity. The transhistorical conception of human identity (*homo faber*: man the tool maker), situated at the core of extension theory, neglects the ways in which technology is both culturally constructed and intricately tied to what it means to be human. Extension theory ties to a cultural conception of the human that can exist intact without the aid of technology. In this view, technology is an 'other' to life.

It follows, then, that the new directions in early 20th-century television depend on a different philosophy of technology. The popular science of television that emerged with full force in the 1920s conveys an inadequacy in the simplistic body-machine metaphor. A popular account of television from 1931 points out the limits to the body-technology analogy. In noting the similarities and differences between telephony and television, the author declares 'it is significant that nature has evolved only a receiving system for visual impressions, and that there is no organism capable of originating visual impressions at will, as we can set up to imitate sound impressions'.[41] The human body's lack of a screen pushes the limits of the metaphor. In cases such as this, the body-machine metaphor reaches the end of its capacity to explain the meaning of technology.

Another writer of popular science begins his explanation of the transmission process by describing the electric eye, a photoelectric cell which constitutes the electronic version of the selenium cell. Instead of simply offering the analogy as shorthand for describing the technical aspects, the author continues by noting the limits to the metaphor. 'Figuratively speaking, this circuit acts as an extensible optic nerve. Unlike an actual nerve channel, it cannot terminate directly in the brain of the observer. Therefore, it terminates in certain electrical equipment—the viewing

39 Brey 7-8.

40 Mitcham, 21-23; Väliaho, 80-82.

41 Edgar H. Felix, *Television: Its Methods and Uses* (New York: McGraw-Hill, 1931), 11.

apparatus.[42] Explanations such as these show how popular science began to incorporate a more sophisticated understanding of the role and meaning of technology in general and television specifically as it affected the process of human visual perception.

Early 20th-century popular science encouraged a new way of thinking about television as a perceptual process, a partnership between humans and technology. This new strategy was represented in a shift in the visual representations of television designs, in the scientific explanation of television systems, as well as in the technical methods adopted to solve the problem of television according to new discoveries in physics.

Metaphors

In the history of technology, metaphors have always provided a strong expression of the meaning of technology. In the nineteenth century, analogies with the telegraph and nervous system established a link between technology and the body. As Nicholas Wade explains, these metaphors have always provided a basis for understanding the meaning and function of technology. The strategy can be found in many different cultures across time.

> Our understanding of perceptual processes[...] has very often been shaped by concepts and models drawn from other fields of scientific enquiry and applied as analogies of the working brain[...]. Such analogies may be widely accepted, to the point of being thought self-evident, but their inadequacy has become [sic] apparent before long[...]. No doubt all such analogies were useful at the time they were proposed, but it is important to be aware that they are speculations rather than explanations, and that this applies as much to the computer as it does to clockwork. It is simply a measure of our ignorance that we do not know how to characterize the operation of the brain in terms that are independent of analogy with other sorts of mechanisms.[43]

Clockwork, automata, hydraulics, telephones, and computers: the correspondence of technology to the body has a long history that extends back long before the Industrial Revolution. Along with the historical analogies of

42 John Mills, *Through Electrical Eyes* (New York: Bell Labs, 1928), 5.

43 Nicholas Wade, *Visual Perception: An Introduction* (New York: Psychology Press, 2013), 23; Carolyn Marvin, *When Old Technologies Were New*, 111; Tom Standage, *The Victorian Internet*, xviii, 211.

clockwork to cognition, automata to the mechanical body, and telegraphy to the nervous system, television came to fill a role at first analogous to the eye and was eventually understood as a mode of perception-at-a-distance. Tracing the long history of these correspondences helps us to recognize the ways in which cultural and intellectual history and the history of technology are imbricated. Culture and technology evolve so intimately that it makes little sense to study them in isolation.

Iwan Rhys Morus argues that technological systems establish body-machine analogies that extend across time.[44] Morus's study of the British telegraph explains how a technological network can become analogous to the human nervous system. He writes: 'The metaphor worked both ways.'[45] If the telegraph network could be understood as operating at the speed of human thought, then the brain could also be understood as a mechanical system. If there is any doubt as to how the metaphor transcended mere visual resemblance, look at how 'telegraph' came to function in informal language to reflect the way the body can unconsciously reveal one's thoughts. The Oxford English Dictionary offers an example: 'a tiny movement of her arm telegraphed her intention to strike'. In a similar way that the telegraph resembled the nervous system, television grew out of the age-old correspondence between camera and eye. Electrical networks revealed a sophisticated metaphor that bound technology with human visual perception.

When Alexander Graham Bell filed his first patent for the telephone, he chose to call it a 'talking telegraph'. Before his success with the telephone, the telegraph had provided the model on which new inventions were based. It encouraged a way of thinking about communicating over a distance that was mediated by a public service, the telegraph office. The telephone broke that mould; it introduced a direct relationship between the user and the machine.

The suggestion that human users had direct access to their distant correspondents established a new way of thinking about communication at a distance. The machine became, in a sense, an extension of the user's natural senses. Hearing at a distance quickly opened up the possibility of *seeing*

44 Iwan Rhys Morus, '"The Nervous System of Britain": Space, Time and the Electric Telegraph in the Victorian Age,' *The British Journal for the History of Science* 33, no. 4 (2000): 455-475. See also Iwan Rhys Morus, ed. *Bodies/machines* (London: Bloomsbury Publishing, 2002); Bruce Mazlish, *The Fourth Discontinuity: The Co-evolution of Humans and Machines* (New Haven, CT: Yale University Press, 1995); David Mindell, *Between Human and Machine: Feedback, Control, and Computing Before Cybernetics* (Baltimore, MD: John Hopkins University Press, 2002); Anson Rabinbach, *The Human Motor: Energy, Fatigue, and the Origins of Modernity* (Berkeley, CA: University of California Press, 1992).

45 Morus, 457.

at a distance. Journalists often remarked how the possibility of seeing by electricity had become an inevitable follow-up to the telephone. The gist of these arguments followed the form: 'Since it has become possible to hear at a distance, why should we not also be able to see?'[46]

The period Stephen Kern refers to as the 'culture of time and space' witnessed a shift in modes of mediated perception, from one based on face-to-face mediation, such as in the telegraph office, to one of machine-mediated communication, as in the user's direct contact with the telephone. With this context in mind, seeing by electricity can be understood as a new way of thinking about time and space. The culture of seeing by electricity drew its expression from the late nineteenth-century conceptions of technological progress and the annihilation of space.

Electronic Television: The Emergence of Systems Thinking

While extension theory characterizes the way seeing by electricity seemed to convey a sense that technology offered users a new ability to 'extend' their reach beyond physical boundaries, a different approach to the design and conceptualization of television arose in the early 20th century. From the direct metaphor of eye-to-camera emerged an extended metaphor of the human visual system, a new relationship between eye-brain cognitive process and electronic television system. I term this transition *the emergence of systems thinking.* Peter Checkland defines systems thinking as 'a particular way of thinking about the world', a model that has existed in varying degrees of popularity since the birth of Western civilization.[47] Checkland places

46 'Seeing by Electricity,' *The Electrician*, 7 March, 1890, 448-450. Statements like this litter the nineteenth-century literature on 'seeing by electricity'. Interestingly enough, once television became a functional technology in the 1920s, this phrase switched to the ability to smell at a distance along with telepathy. See, for example, Moseley's extravagant claims in the introduction to his book, *Television: Today and Tomorrow* (London: Sir Isaac Pitman & Sons, 1934).

47 Peter Checkland, *Systems Thinking, Systems Practices* (New York: John Wiley, 1981), 3. According to Checkland, systems thinking relies on the analogies formed between organic (living) and technical systems or processes. Most works on 20th-century systems thinking focus on the rise of systems theory and cybernetics. Any connection between early 20th-century developments in television and what is conventionally understood as 20th-century systems theory only becomes clear when placed in the context of Bell Labs's second iteration of the two-way television project in the 1960s. Its lead engineer authored a textbook recognized as a key introduction to hard systems theory. See A.D. Hall, *A methodology for systems engineering* (New York: Van Nostrand, 1962); AD Hall, 'Experiments with Picturephone Service,' *Bell Labs Record* 42 (April 1964): 114-120; John Mingers, *Realising Systems Thinking: Knowledge and Action in Management Science* (New York: Springer Science & Business Media, 2006), 1. On 20th-century

particular emphasis on the way systems thinking draws inspiration from organic (living) models. However, as becomes apparent in his survey of systems thinking across the centuries, each cultural historical moment carries with it a particular social agenda.

According to the applications of systems thinking to the problem of television in the early 20th century, the process of human seeing provided the model for how television systems should function.[48] The metaphors persisted, though emphasis shifted *from organs to systems*. The nineteenth-century notion of extending the range of vision gave way to a bio-technical construction of the human visual system analogous to the electronic television apparatus. Television evolved as a system analogous to the telephone and electric power.[49] Just as the telegraph crossed the nation like a nervous system, television grew into a functional metaphor for the human *visual system*. The shift can be seen represented both in the scientific and popular discourses as well as in their associated methods of illustration and design.

One early 20th-century work of popular science, *Modern Inventions* (1915), describes 'The Human Eye as a Model':

> Optically speaking, the eye is a camera obscura containing a lens by means of which the image of what is looked at is cast upon the retina, as on the focusing screen of an ordinary camera. The surface of the retina is connected through the optic nerve with the brain by means of a very large number of little threads or nerve fibers, each of which is joined to a certain definite point on the retina, and which when stimulated by the action of the electro-magnet waves which we term little communicates to

systems theory, see Ludwig Von Bertalanffy, 'The History and Status of General Systems Theory,' *The Academy of Management Journal* 15, no. 4 (Dec 1972): 407-426; F. Emery, ed. *Systems Thinking* (New York: Penguin, 1969); Darrell Arnold, ed. *Traditions of Systems Theory: Major Figures and Contemporary Developments* (New York: Routledge, 2013).

48 Thomas P. Hughes, 'The Evolution of Large Technological Systems,' in *The Social Construction of Technological Systems: New Directions in the Sociology and History of Technology*, eds. Trevor Pinch, Thomas Hughes, and Wiebe Bijker (Cambridge, MA: MIT Press, 2012; 1987); Erik van der Vleuten, 'Large Technical Systems,' in *A Companion to the Philosophy of Technology*, eds. Jan Olsen, Stig Andur Pedersen, and Vincent F. Hendricks (New York: John Wiley & Sons, 2012).

49 In the history and philosophy of technology, the study of large technical systems (LTS) constitutes a distinct approach of its own, attributed to the sociology and history of science that became popular in the 1980s along with the social construction of technology (SCOT) and actor-network theory (ANT). Hughes, 'The Evolution of Large Technological Systems'; van der Vleuten, 'Large Technical Systems'.

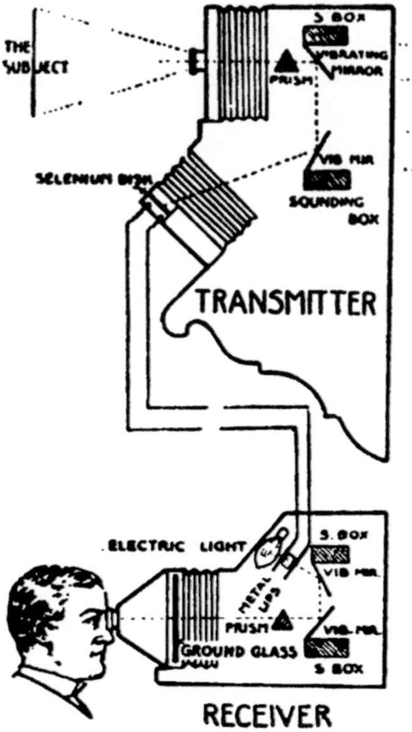

Cross-section of the receiving and transmitting boxes of the Telectroscope.

Figure 27. Szczepanik's Telectroscope. Cleveland Moffett, 'Seeing by Electricity,' *Pearson's Magazine*, 1899.

> the brain, in a mosaic form, an idea or conception of the various portions of the image.[50]

This writer's description suggests that the simple 1:1 analogy between eye and camera no longer provided an adequate model for television. As television developed into a system, it became necessary to extend the analogy to include the process of human vision. While the description still begins with noting the foundation of the technology in the eye as a model, it develops into a more sophisticated analogy with the human visual system.

As electronics and the systems approach became a general model, engineers moved away from the design of individual components, which

50 V. E. Johnson, *Modern Inventions* (New York: FA Stokes, 1915), 241.

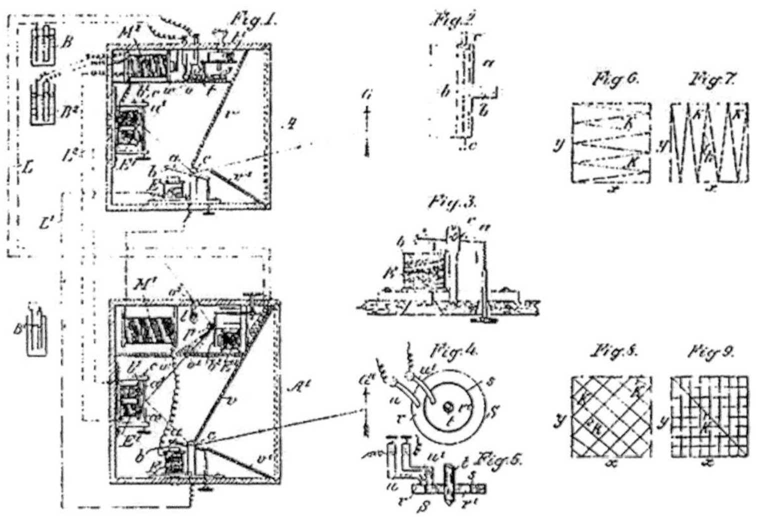

Figure 28. Jan Szczepanik, '*Method and Apparatus for Reproducing Pictures and the like at a Distance by Means of Electricity*, British Patent 5031, 1897.

supported a way of thinking about the media as adjunct, to the senses, to a more sophisticated model of the technology-systems approach, which supports a way of thinking about humans and technology as symbiotic. Illustrating this concept, television inventor John Logie Baird used the term 'human television system' in one *New York Times* interview, describing it as 'an apparatus in imitation of the human optical system. The human eye consists essentially of a lens which casts an image of the object viewed upon the retina'.[51] A new concept of vision and tele-vision emerged along with the notion of television as a system. It integrated the process of human vision, and, in the literature, it became more common to refer to vision as a system.[52] The 'Human visual system' incorporated the eyes, retina, and brain while placing a new emphasis on the process by which the brain made sense of the image.

The emergence of systems thinking can be discerned as early as 1899, with Cleveland Moffett's illustration of Szczepanik's telectroscope that

51 'Glasgow Listens to Sound of Faces,' *New York Times*, 4 February 1927.

52 J. H. Nelson, 'Ideal Seeing Conditions: The Study of the Human Visual System as a Basis for Prescribing Lighting,' *British Journal of Industrial Medicine* 2, no. 4 (1945): 224; Leon Harmon, 'Analogs and Models of the Human Visual System,' *Optometry & Vision Science* 36, no. 6 (1959): 304-312; Douglas Granrath, 'The Role of Human Visual Models in Image Processing,' *Proceedings of the IEEE* 69, no. 5 (1981): 552-561; Dorothea Jameson and Leo Hurvich, eds., *Visual Psychophysics* (New York: Springer, 1972).

was published in the American periodical *Pearson's magazine* (Figure 27).[53] It pictures a generic human observer, a man in profile, whose eye is placed in immediate contact with a metallic viewer. The figure offers the suggestion of 'the subject' at the other end of the line. This 'cross-section of the receiving and transmitting boxes' suggests a system altogether different from the machines that existed before it. Szczepanik's telectroscope shows a system, no longer just a stand-alone camera or screen, which takes into account the human observer and his television 'subject'.

This case offers a unique overlap of what seems like two different worlds. Compare Moffett's systems illustration, for example, with two other approaches to the representation of the telectroscope. One, from the patent, emphasizes the process by which the system was designed according to the principles of electrical engineering (Figure 28). Another, from the *Illustrated London News*, emphasizes the telectroscope in a way that would be recognizable to a popular reader (Figure 24). It shows how the system uses mirrors to transmit an image from one place to another. Looking at these images side by side reveals three different communities of thought. The technical diagram establishes the legitimate scientific nature of the invention by communicating the function of the device according to the conventions of electrical engineering. The *Illustrated London News* follows a traditional Victorian approach according to the culture of seeing by electricity, which emphasized the heroic character of the inventor, but with the slight difference of appealing to an audience versed in theatrical stage illusions. Moffett's systems diagram would have appealed to general readers of the periodical, while also representing a new scientific bent in the development of television technology. Representing the telectroscope as a system rather than a singular camera or screen marks a departure from the established culture of seeing by electricity. It suggests that the way of thinking about television as a system linking two places together had begun to take hold in the technical community as well as in the popular culture.

It wasn't until 1908, when physicist A. A. Campbell Swinton proposed applying scientific methodology to the 'problem of television', with the use of cathode rays, that a change in the culture of seeing by electricity became apparent.[54] As a physicist, Swinton looked on the problem from an altogether

53 Cleveland Moffett, 'Seeing by Electricity,' *Pearson's Magazine*, October 1899, 493.

54 A. A. Campbell Swinton, 'Distant Electric Vision,' *Times*, 15 November 1911; A. A. Campbell Swinton, 'The Possibilities of Television with Wire and Wireless,' *Wireless World*, 9 April 1924, 51-56; 16 April 82-84; 23 April 114-118; A. A. Campbell Swinton, 'Electric Television,' *Nature* 118, no. 2973 (October 1926), 590.

different direction than those who had tackled it in the past. Harnessing the power of the electron opened up new possibilities. Swinton and his colleague Silvanus Thompson described television as at once a simple concept and a complex technical problem.[55] The central idea of seeing by electricity seems simple enough. But to achieve a practical result, they explained, required a sophisticated knowledge of and mastery over the physical world. Swinton was not the first to suggest using electronics, but his respected position in the American scientific community and the confidence with which he spoke gave his message the force needed to generate interest in a new direction for television, which he referred to as 'distant electric vision'.[56] Vacuum tubes and electron beams transformed television into a technology so scientifically complex that amateurs no longer played a role.

His contribution reached the scientific community in the form of a letter published in *Nature* in response to a statement made by Shelford Bidwell. Bidwell had written, in frustration, about the barriers halting progress in telegraphic photography, principally the 'sluggishness' of the selenium element. Swinton responded: 'it is wildly impracticable to effect even 160,000 synchronized operations per second by ordinary mechanical means'.[57] Coming at the 'problem of television' from the world of physics and applied science made Swinton's contribution new and noteworthy. His approach differed from the established methods for 'seeing by electricity' and encouraged a new way of thinking about television as a scientific endeavour. He gave voice to a new generation of electronic engineers, and to a new approach to television.

Swinton's expression of television, encapsulated in 'distant electric vision', offers insight into the way the scientific community drew on the expectations established by 'seeing by electricity'. The culture of telectroscopes had been translated through the language of science, only to make its way back into the popular culture of the teens. Swinton opened his 1912 presidential address to the Röntgen society, for example, with a familiar exclamation about man's mastery over nature and scientific power.[58] But, to this general concept, he adds several new features. First, he privileges science over

55 A. A. Campbell Swinton, 'Presidential Address (with concluding remarks by Silvanus Thompson),' *Journal of the Röntgen Society* 8, no. 30 (1912): 8, 15.

56 Robert Grimshaw, 'The "Telegraphic Eye",' *Scientific American*, 1 April 1911, 335-336; Prof. Rozing's 'Electric Eye' – A New Apparatus for Television,' *Scientific American Supplement*, 71, no. 1850 (June 1911): 334.

57 Shelford Bidwell, 'Telegraphic Photography and Electric Vision,' *Nature* 78, no. 2014 (1908): 105-106; Swinton, 'Distant Electric Vision,' 151.

58 Swinton, 'Presidential Address,' 1.

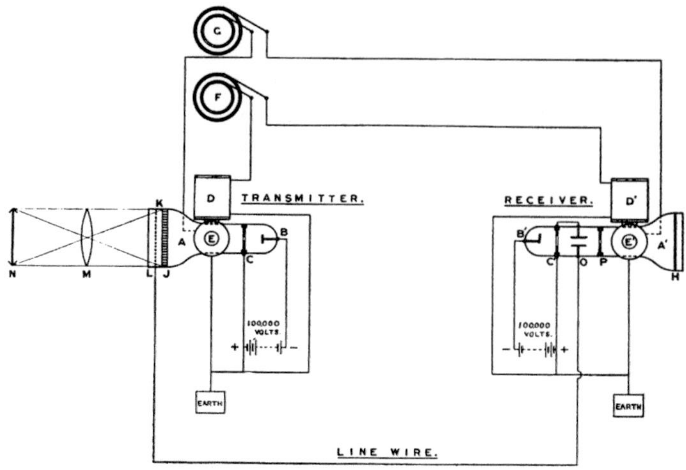

Figure 29. A. A. Campbell Swinton, schematic for distant electric vision, 'Presidential Address,' *Journal of the Röntgen Society* 8.30, 1912, 10.

Figure 30. Hugo Gernsback, 'Television and the Telephot,' *Electrical Experimenter*, May 1918.

technology, and scientific practice over practical invention. Approaching an old concept from a new perspective, Swinton describes seeing by electricity as a problem to be solved by science: 'It supposes an entirely new application of Crookes tubes and the phenomena of Cathode Rays.' Beginning with the science, Swinton then introduces the problem to be solved with this new knowledge and practice: 'distant electric vision, or the power to see objects a great way off by electrical means[...] [an] extension of our sense of vision'.

Second, he extends the body-machine metaphor into a systems analogy. Though introduced through the simple metaphor, his description emphasizes visual processes and the perception of images in the brain analogous to the function of a television system. In addition to recognizing the human eye as a model for distant vision, just as the ear modelled the telephone, Swinton develops an extended analogy that draws on knowledge of the visual processes involved in perceiving images in the brain.[59] Far from the simple 1:1 eye-camera analogy, 'distant electric vision' articulates a more complex interpretation of the human visual system applied to the problem of seeing by electricity.

A look at the visual representations of 'distant electric vision' illustrates how the new approach to television made its way into American popular culture and popular science. While Swinton presented the scientific community with a technical schematic (Figure 29), the popular press experimented with new ways of illustrating the concept to the public. These illustrations gave the image of television a heightened appeal and introduced a new way of representing television. A notable example can be found in Hugo Gernsback's popular-science periodical *Electrical Experimenter.* The magazine was marketed to and targeted practical-minded hobbyists, like Gernsback's earlier publishing venture *Modern Electrics*. Both mingled radio news and 'wider aspects of scientific experimentation' with a Q&A section and how-to articles.[60] But unlike its predecessor, *Electrical Experimenter* included more illustrations, a larger format, and short fiction in its colourful pages. The magazine combined the attractiveness of pulp with a practical approach

59 Swinton, 'Presidential Address,' 7; A. A. Campbell Swinton, 'The Possibilities of Television with Wire and Wireless,' *Wireless World*, 9 April 1924, 51.

60 Michael Ashley and Robert Lowndes, *The Gernsback Days: A Study of the Evolution of Modern Science Fiction from 1911 to 1936*, (Holicong, PA: Wildside Press LLC, 2004), 33-34. See also Keith Massie and Stephen D. Perry, 'Hugo Gernsback and Radio Magazines: An Influential Intersection in Broadcast History.' *J. Radio Stud.* 9 (2002): 264; David E. Sumner, *The Magazine Century: American Magazines Since 1900* (Bern: Peter Lang, 2010); Grant Wythoff, 'Pocket Wireless and the Shape of Media to Come, 1899–1922,' *Grey Room* 51 (2013): 40-63.

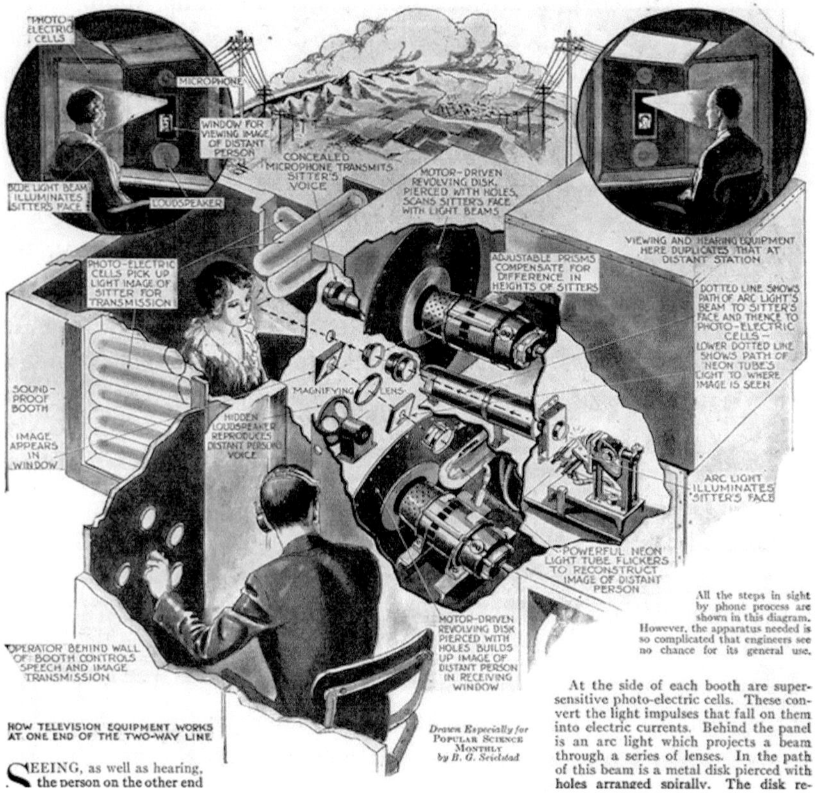

Figure 31. 'How Television Equipment Works at One End of the Two-Way Line,' illustrated by H. G. Seidstad for *Popular Science*. 'Talk, Hear, SEE on This Phone: Two-Way Television Is Demonstrated in Laboratory As an Engineering Stunt,' *Popular Science*, July 1930, 22.

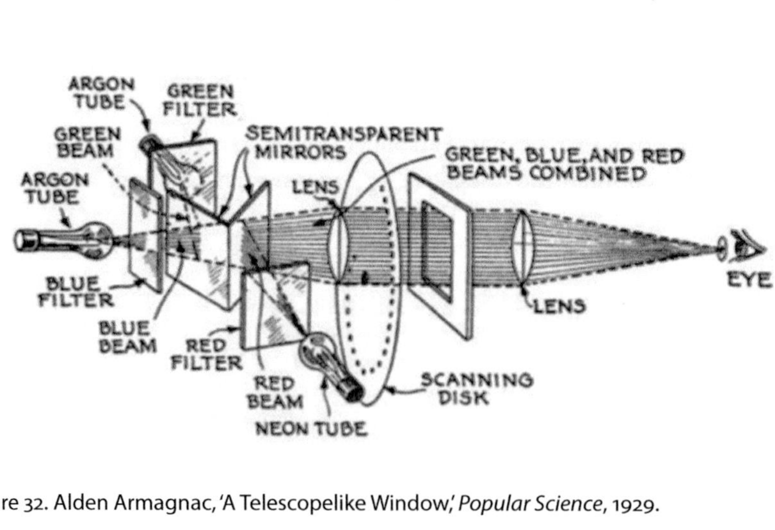

Figure 32. Alden Armagnac, 'A Telescopelike Window,' *Popular Science*, 1929.

Figure 33. Illustration by Will B. Johnstone, 'Face with the Smile,' *New York Morning World*, 10 April 1930.

Figure 34. H. W. Secor, 'Television, of the projection of pictures over a wire,' *Electrical Experimenter* 3, August 1915, 131-132.

to science and technology, fostering the emergence of mainstream science fiction and popular science.

While the culture of seeing by electricity pictured screens like mirrors or looking glasses, Machine-Age television developed according to a similar aesthetic model. Hugo Gernsback imagined his Telephot as a handheld mirror (Figure 30). The Bell Labs Two-Way television seemed to offer a 'window for viewing [an] image of [a] distant person' (Figure 31). One writer of popular science referred to electronic television as a 'telescopelike window' (Figure 32).

Fears of technological change also persisted along similar lines. Journalists echoed the nineteenth-century anxieties of surveillance and privacy articulated in the press response to the far-sight machine in the form of sarcastic commentary. When Bell Labs unveiled their Ikonophone in 1930, it gave concrete form to those fears, inspiring one cartoonist to picture the potential for telephone users to spy on women while they were in the shower (Figure 33). In fact, fears of the invasion of privacy never completely went away and, in some respects, were heightened when the prospect of video telephony re-emerged later in the 20th century. One journalist offered a similar critique of the Picturephone when Bell brought the idea back in the 1950s. The illustrator's depiction of the astonished telephone user assaulted by the bill collector bears such close resemblance to 'Professor Goaheadison's Latest' and the criticisms against Edison's 'far-sight machine' from 1889 that it supports, somewhat deceptively, a sense that the idea of television has remained relatively stable over time.[61]

Electrical Experimenter's article on Swinton's 'distant electric vision' (1915) features two distinct approaches to the representation of television (Figure 34).[62] While the discussion and representation of 'seeing by electricity' had tended to be directed towards either scientific or popular readership, *Electrical Experimenter* departs from that convention by combining both approaches. The first consists of two figures that make up a banner across the head of the article. Illustrating the television user's perspective, it identifies the 'man at right being transmitted and reproduced on screen in front of lady. Her face is transmitted and reproduced [...] before man'. Between the two distant correspondents stretch electrical lines that extend across a

61 John Gould, 'Picture, Please!: How You'll be Able to See While you Talk on the Newest Version of Mr. Bell's invention,' *New York Times*, 16 September 1956; 'Professor Goaheadison's Latest,' *Fun*, 3 July 1889: 6; 'Goaheadison's Real Latest,' *Fun*, 17 July 1889: 24; 'Untitled' (Edison's Far-Sight Machine), *Baltimore Herald*, 10 June 1889; 'Untitled' (Edison's Far-Sight Machine), *Lincoln Journal* (Nebraska), 26 May 1889.

62 H. Winfield Secor, 'Television, Of the Projection of Pictures Over a Wire,' *Electrical Experimenter* (August 1915): 131-132; 172-174.

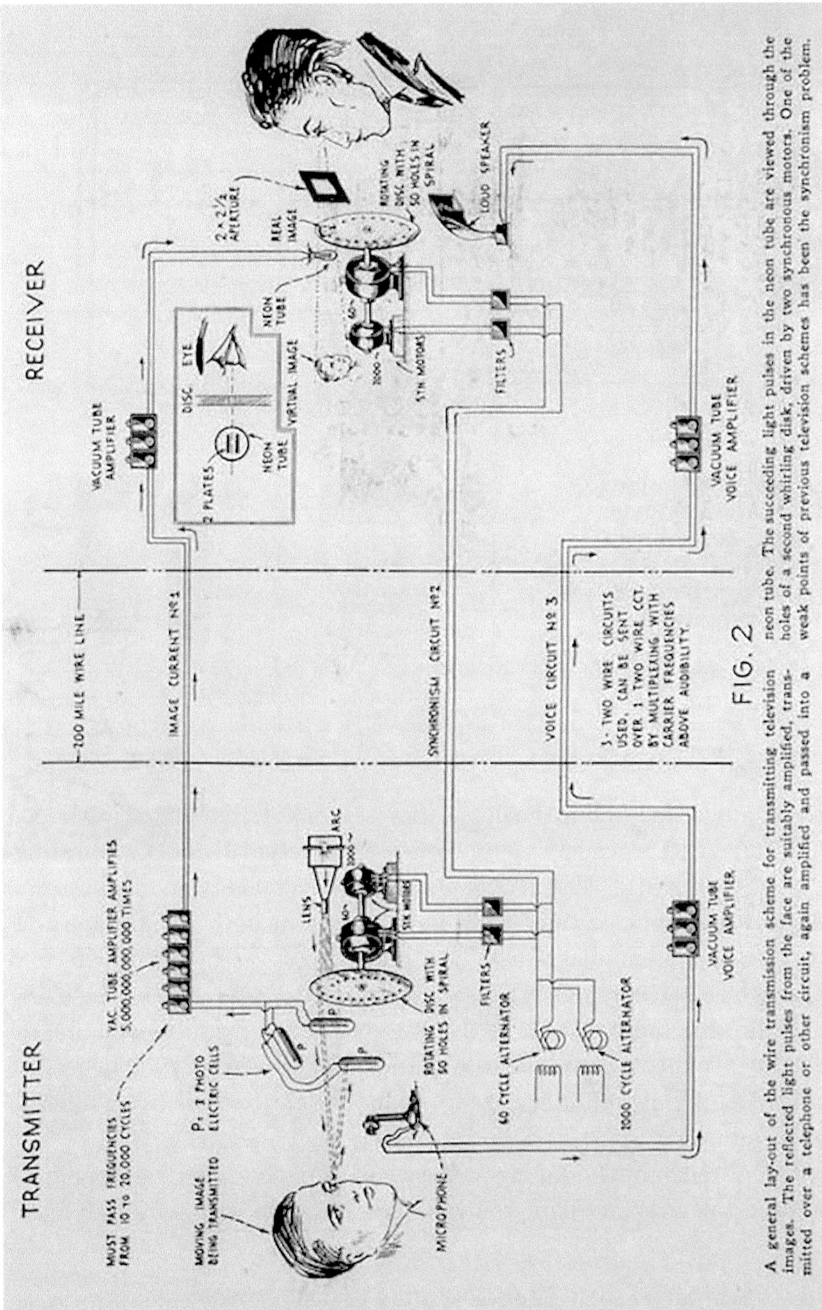

Figure 35. H. W. Secor, 'Television Perfected at Last,' *Science and Invention*, 1927.

Figure 36. C. H. W. Nason, 'A Short Course in Television,' *Television News*, 1931, 31.

pastoral landscape. While the illustration resembles a cinematic montage, it also draws on the reader's appreciation of the actual distance separating the correspondents.[63] The article also reproduces a version of Swinton's technical schematic on the reverse page. Including both a landscape and an electronic-systems diagram in the same article suggests that the concept of seeing by electricity was in a state of transition. Engineers and the reading public alike were witness to the merging of two perspectives: that of the cultural expectations of distance and communication established by seeing by electricity and the new frontier of scientific possibility offered by electronics.

The figures published in Secor's article provide an early example of the trend towards picturing television as a system with a human user.

63 Jan Olsson, 'Framing Silent Calls: Coming to Cinematographic Terms with Telephony,' in *Allegories of Communication: Intermedial Concerns from Cinema to the Digital*, eds. John Fullerton and Jan Olsson (Bloomington, IN: Indiana University Press, 2004), 157-192.

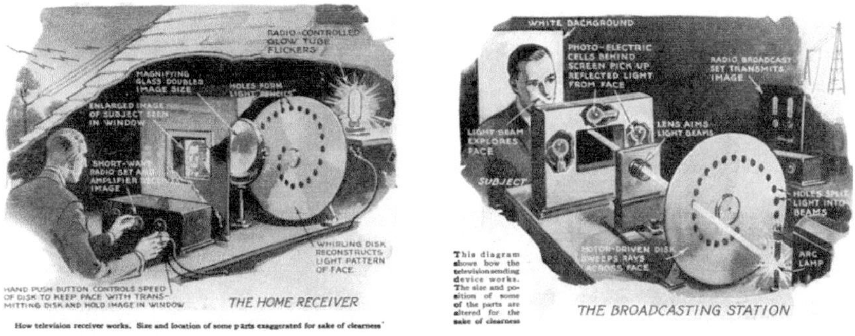

Figure 37. Armagnac, 'Television Brought into the Home,' *Popular Science Monthly,* April 1928, 20-21, 143.

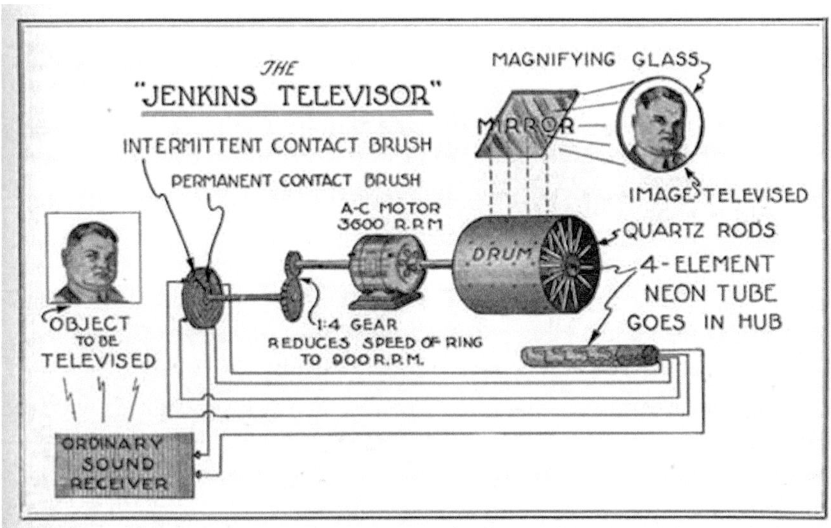

Figure 38. Jenkins Televisor, *Modern Mechanics,* January 1929.

Continuing the tendency to intermingle popular and technical illustrations, Figures 35-38 document a new mode of picturing television as such a system. In popular science periodicals like *Science and Invention, Modern Mechanics,* and *Popular Science,* artists adopted an approach that fused the technical appearance of schematics with the visual rhetoric of nineteenth-century television. Whereas Figure 34 represents two different strategies in representing television, those strategies appeared to merge into one another as popular-science magazines placed more importance on illustrations in the 1920s. These diagrams illustrate how the technological system provides a means to mediate the relationship between the human and the screen.

One example comes from a 1927 article of Secor's: 'A general lay-out of the wire transmission scheme for transmitting television images' (Figure 35).[64] This early attempt to illustrate Bell Labs's two-way television system shows care taken in characterizing the user. Identified in Figures 31, 39-41 as observers, subjects, or simply pictured as an eye, these pictures show how the user became *a part of the system.* These illustrations consistently depict the user's eye connected to the apparatus using a dashed line. Travelling from the user to the lamp, screen, or disc, it emphasizes the way the user *sees into or through* the machine. Connecting the eye to the apparatus in this way can be understood as a strategy for engineering as well as a recognizable reference to the viewer's line of sight.

'A general lay-out' connects two people across an abstract length of space: '200-mile wire line'. Suggesting an expanse of space, it represents the connection between two distant correspondents over a *virtual* divide. Compared to the representation of a natural landscape from the 1915 depiction, for example, it marks a shift from representing television as the latest in space annihilation to the process of technologically mediated vision. As it was no longer possible to depict the immensity of the television system as it extended across space, representations shifted towards an emphasis on the user's process of vision and the mode of electrical mediation. A single electrical wire would no longer suffice. Space became an abstract concept, as did the perceived distance separating the users.

A similar illustration appeared in *Television News* (1931), along with the description: 'Approximate representations (in graphic form) of what goes on in a complete television system)′ (Figure 40).[65] Like 'a general lay-out', Nason's 'approximate representations' picture two users looking through an apparatus. It follows the strategy seen in Secor's representation of the line of sight. The users appear in strict profile, emphasizing the way they seem to be looking at each other. In contrast, Secor's subject 'being transmitted' at left appears at a slight angle making it seem like his gaze is directed at the 'lens' rather than at the viewer. At the other end of the line, Secor depicts the viewer at right gazing at a miniaturized reproduction with an attentive look on his face. Simplified from the detail offered in Secor, Nason's 'image current',

64 H. W. Secor, 'Television Perfected at Last,' *Science and Invention* (June 1927): 108, reprinted in Alfred Dinsdale, *First principles in Television* (New York: Wiley, 1932): 88.

65 C. H, W, Nason, 'A Short Course in Television,' *Television News* 1, no. 1 (March-April 1931): 30-31, 71.

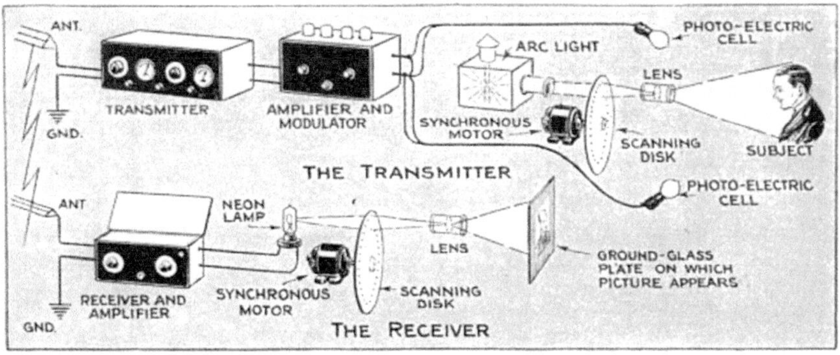

Figure 39. 'How the Light Waves from the Subject are Transmitted by Radio and Converted back into a Picture at the Receiver,' in 'What Television Offers You', *Popular Mechanics*, November 1928, 820-824.

Figure 40. Sanabria Television System, *Television News*, September-October 1931.

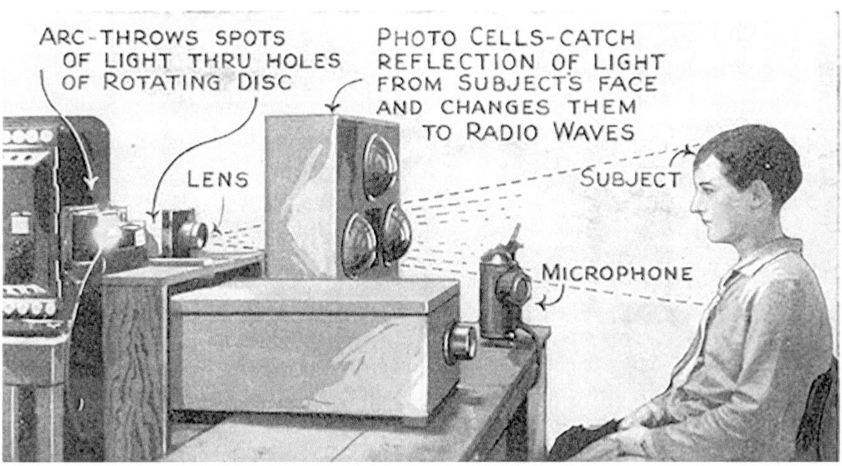

Figure 41. 'The Transmitter for Radio Pictures,' in 'Television for the Home,' *Popular Mechanics*, April 1928, 531.

depicted as an electrical cable, draws a direct connection between the eyes. The arrows travelling along Nason's electrical cable, more recognizable in the simplified composition than in Secor's depiction of both image and voice channels, emphasizes the directionality of the signal moving from the onscreen subject at left to the television observer at right across an abstract distance by means of an electronic signal. Replacing the enumeration of the many components that make up the apparatus depicted in Secor's 'general lay-out', with an abstract representation of electrical signals allows Nason's 'approximate representations' to show how the electrical signal mediates the vision between two users. It departs from the depiction of an actual or virtual landscape in order to emphasize the electronic mediation of vision.

Scientific illustrations continued to depict the line of sight with dashed lines. This also became a common strategy in human engineering (now known as human-factors engineering or ergonomics) as a way of portraying the user's interface with a machine (human-screen interaction).[66] The use of human-engineering strategies in television design is most apparent in the way Bell Labs depicted their two-way television project.[67] As represented by Bell Labs's engineers, 'a pictorial sketch of two-way television system' features the viewer's full body and a chair. Like 'a general lay-out' and 'approximate representations', 'pictorial sketch' features an observer with lines emanating from his eyes. Comparing such figured with schematics from ergonomics illustrates how human-screen interaction informed the design of television systems in the 1920s.

There is a marked contrast between the visual rhetoric of seeing by electricity with the depictions of television systems that became popular in the 1920s. While the former provide a visual representation of the annihilation of space, the latter depict television as a system with a human user. These illustrations privilege the user's interaction with an interface. Similar to the way the jagged and dotted lines in the systems diagrams visualized the abstractness of mediation, these pictures show how the image represented

66 For a critical perspective on human engineering, see Donna Haraway, *Primate Visions: Gender, Race, and Nature in the World of Modern Science* (New York: Psychology Press, 1989), 66; Donna Haraway, *Simians, Cyborgs, and Women: The Reinvention of Nature* (New York: Routledge, 2013), 11, 22, 47. On human-factors engineering and ergonomics, see Ernest McCormick, *Human Factors Engineering* (New York: McGraw-Hill, 1964); Christopher Nemeth, *Human Factors Methods for Design: Making Systems Human-centered* (Boca Raton, FL: CRC press, 2004); Stephen J. Guastello, *Human Factors Engineering and Ergonomics: A Systems Approach* (Boca Raton, FL: CRC Press, 2013).

67 *First Demonstration of Two Way Television: A Development of the Bell System* (New York: Bell Labs, 1930).

on the screen will function as the focal point of televisual mediation. The prominence of the observer in these depictions emphasizes how integral human visual perception had become to the operation of television. If distant electric vision would be possible, it would be a matter of learning how the human fit into the machine.

Taking into account the emergence of systems thinking in the development of television in the 1920s requires a media theory that recognizes how the user's visual perception is mediated by the technological apparatus. A theory of technological extensions focusses primarily on the material, mechanical, and physical properties that link the body to the machine. In its nineteenth-century formulation, extension theory presents a weak conception of the intrinsic hybridity of human and machine.

In order to recognize how deeply entangled the human and the machine are, we should turn to other approaches that define human and machine more broadly. Mark Hansen's media theory, for example, understands technology as a central factor in the human life environment, along with culture and visual representation. His approach considers a redefinition of media and technology from the perspective of embodied perception. In a work co-authored with visual-studies scholar W. J. T. Mitchell, he seeks to distance 'media' from a conception based in representations (sounds and images) to one based on embodied perceptions. In addition, he replaces *technology* with *technics*, 'a practical knowledge emanating from skill, art, or practice', which allows for an expanded interpretation of the role of technology in culture distanced from a foundation in mechanical artefacts.[68] They write: '*media*, in our view, also names a technical form or formal *technics*, indeed a general mediality that is constitutive of the human as a "biotechnical" form of life.'[69] Used as an alternative to the philosophy of technology, media theory breaks down artificial divisions between technics and culture in order to treat them both as aspects of a lived environment.

Elsewhere, Hansen extends this definition into a theory for understanding '*medium as environment for life*'.[70] 'Media theory' reminds us that there is no such thing as 'unmediated' perception, just as culture and nature work together to construct a sense of 'reality'. From this perspective, *media* in general and *television* in particular define a way of seeing as a kind of visual

68 Mark Coté, 'Technics and the Human Sensorium: Rethinking Media Theory Through the Body,' *Theory & Event* 13, no. 4 (2010): n.p.

69 W. J. T. Mitchell and Mark B. N. Hansen, eds. *Critical Terms for Media Studies* (Chicago, IL: University of Chicago Press, 2010), ix (my italics).

70 Mark B. N. Hansen, 'Media Theory,' *Theory, Culture & Society* 23, no. 2-3 (2006): 299-300.

perception that is a culturally and historically specific learned behaviour. *'Medium as environment for life'* suggests that moving-image technology and human visual perception co-exist and co-evolve. The emerging complexity suggests that visual media and visual perception are inextricably connected. Our way of seeing the world constructs our way of representing it, and vice versa.

Conclusion

In comparison to television's nineteenth-century 'speculative era', which encompasses the culture of seeing by electricity, a new mode of representation emerged in the early 20th century. Along with scientific developments in electronics, systems thinking came to dominate both the technical and cultural modes of representing what television was and what it would become. Along with the introduction of these scientific discoveries emerged a modern construction of vision that relied on the concept of technological mediation.

Both 'seeing by electricity' and 'distant electric vision' comprise television's speculative era. While both appear to resemble television in form and function, there are several important differences that should be recognized. The metaphor of vision no longer drew a direct connection between the eye and the televisual mechanism. While the metaphor of 'the human eye as a model' persisted, many writers of popular science were quick to make clear how the metaphor provided merely a figurative correlation. As a sign that the hyperbolic claims of the annihilation of space had finally come true, representations of television in the Machine Age emphasize the picture on the screen as well as the screen itself as a mediating device. The image on the screen perceived by the viewer comes to seem more and more separated from its real-world referent. The representations, underlying science, and discourse of television changed significantly enough between the nineteenth and 20th centuries to constitute a rebirth of the medium.

But, in essence, its form and function remained relatively unchanged. Television continued to be described as a tool to connect people across great distances in real time. Whether a one-way relay of the images and sounds of a theatrical entertainment or a two-way communications medium, the general functions remained consistent. The appearance of television also remained stable.

'Seeing by electricity' materialized in both cultural and technological circles. While, as a technology, it clearly resembles the modern concept of television, its visual rhetoric differs dramatically. The nineteenth-century culture of seeing by electricity conveys a sense of unmediated vision. Discussion of annihilating space, communicating with friends and family at a distance, or witnessing a live theatrical performance makes no mention of the way the technology itself forms a barrier to the feeling of presence-at-a-distance. It established expectations of direct access, lacking a sense of mediation or representation. Telectroscopes and telephonoscopes were magic mirrors infused with the real possibilities facilitated by technological achievements. In many respects, the telectroscope was not a technology through which to see. Rather, it represented a culture expecting to have the world at its fingertips. Seeing by electricity functioned as the visual representation of the annihilation of space.

While the desire to annihilate space and the enthusiasm for technological progress had driven the culture of seeing by electricity, the Machine Age infused television with new possibilities fuelled by scientific and industrial mastery over nature. For all intents and purposes, distant electric vision was a different beast entirely. Distant electric vision carries with it assumptions about the limitations of the body and mind in a physical world, infusing electronic television with a sense of control over nature and manipulation of visual perception. Electronics, systems, and the efficiency movement carry with them new connotations for Machine-Age television. In this period, the visual culture of science and technology rewrote the agenda as one of engineering vision, rather than establishing a sense of closeness to those far away.[71]

The next chapter picks up the story in the 1920s, when electronic television re-emerged in popular culture after a hiatus in the physical laboratory. Swinton's promotion of using science to inform technological developments inspired a new generation of engineers. Over the next decade, television was hidden away in the physical laboratory. When it re-emerged in the 1920s, it had transformed into *a new kind of seeing-technology*. New processes and practices, including illuminating engineering, photometry, colorimetry, psychophysics, and the philosophy of the efficiency movement, facilitated the rebirth of the medium.

71 For example, historian of technology Lewis Mumford describes the Machine Age in his 1933 *Technics and Civilization* as a culture between two worlds. He borrows a geological metaphor to describe how technologies can change while seeming to stay the same in different cultural moments: the pseudomorph. See Lewis Mumford, *Technics and Civilization* (Chicago, IL: University of Chicago Press, 2010 [1934]), 265.

A new understanding of electronically mediated communication came to dominate the discourse in the 1920s. In both scientific and popular periodicals, the problem of television became a matter of engineering both the technology and the human visual perception of the image. Technicians became more likely to describe television as a screen interaction distinguishable from face-to-face interaction. When television re-emerged in American popular culture in the 1920s, a host of assumptions, scientific methods, and technical practices came along with it, which proved difficult to explain to the public. A new language of popular science would mediate the technical concepts and terminology for the benefit of the consumer.

Bibliography

Andriopoulos, Stefan. *Ghostly Apparitions: German Idealism, the Gothic Novel, and Optical Media*. New York: Zone Books, 2013.

Andriopoulos, Stefan. 'Psychic Television.' *Critical Inquiry* 31, no. 3 (2005): 618-637.

Ashley, Michael and Robert Lowndes. *The Gernsback Days: A Study of the Evolution of Modern Science Fiction from 1911 to 1936*. Holicong, PA: Wildside Press LLC, 2004.

August, E., 'Seeing by Electricity,' letter to the editor, *English Mechanic*, July 3, 1891, 409.

Baltimore Herald. 'Untitled (Edison's Far-Sight Machine).' June 10, 1889.

Bidwell, Shelford. 'Telegraphic photography.' *Journal of the Society of Telegraph Engineers and Electricians* 10, no. 38 (September 1881): 354-360.

Bidwell, Shelford. 'Telegraphic Photography and Electric Vision.' *Nature* 78, no. 2014 (June 1908): 105.

Bidwell, Shelford. 'Telephotography.' *Nature* 23 no. 589 (February 1881): 344.

Boston Daily Advertiser. 'The Far-Seer.' March 21, 1898.

Campbell-Swinton, A.A. 'Distant Electric Vision,' letter to the editor, *Nature* 78, no. 2016 (June 1908): 151.

Brey, Philip. 'Technology as Extension of Human Faculties.' In *Metaphysics, Epistemology and Technology,* edited by Carl Mitcham, 59-78. London: Elsevier/JAI Press, 2000.

Burns, R.W. *Television: An International History of the Formative Years.* London: Institutions of Electrical Engineers, 1998.

Campbell-Swinton, A.A. 'Distant Electric Vision,' letter to the editor, *Nature* 78, no. 2016 (June 1908): 151.

Campbell-Swinton, A.A. 'Electric Television.' *Nature* 118, no. 2973 (October 1926): 590.

Campbell-Swinton, A.A. 'The Possibilities of Television with wire and wireless.' *Wireless World.* April 9, 1924; April 16, 1924; April 23, 1924.

Campbell-Swinton, A.A. 'Presidential Address (with concluding remarks by Silvanus Thompson).' *Journal of the Röntgen Society.* January 1912, 1-15.

Checkland, Peter. *Systems Thinking, Systems Practices.* New York: John Wiley, 1981.

Coté, Mark. 'Technics and the Human Sensorium: Rethinking Media Theory Through the Body.' *Theory & Event* 13, no. 4 (2010).

Crary, Jonathan. *Techniques of the Observer: On Vision and Modernity in the Nineteenth Century.* Cambridge, MA: MIT Press, 1992.

English Mechanic. 'The Telectroscope.' February 11, 1881.

The Electrician. 'Seeing by electricity.' March 7, 1890, 448-450.

Felix, Edgar H. *Television: Its Methods and Uses.* New York: McGraw-Hill, 1931.

Fickers, Andreas. 'Television.' In *The Handbook of Communication History,* edited by Peter Simonson, Janice Peck, Robert T. Craig, and John Jackson, 239-255. New York: Routledge, 2013.

First Demonstration of Two Way Television: A Development of the Bell System (New York: Bell Labs, 1930).

Fun. 'Professor Goaheadison's Latest.' July 3, 1889.

Fun. 'Goaheadison's Real Latest.' July 17, 1889.

Galili, Doron. 'Seeing by Electricity: The Emergence of Television and the Modern Mediascape, 1878-1939.' PhD diss., University of Chicago, 2011.

Gaycken, Oliver. "The Swarming of Life': Moving Images, Education, and Views through the Microscope.' *Science in Context* 24.03 (2011): 361-380.

Gould, John. 'Picture, Please!: How you'll be able to see while you talk on the newest version of Mr. Bell's invention.' *New York Times.* September 16, 1956.

Granrath, Douglas. 'The Role of Human Visual Models in Image Processing.' *Proceedings of the IEEE* 69 (May 1981): 552-561.

Grimshaw, Robert. 'The 'Telegraphic Eye'.' *Scientific American.* April 1, 1911.

Gunning, Tom. 'Re-newing Old Technologies: Astonishment, Second Nature, and the Uncanny in Technology from the Previous Turn-Of-The-Century.' In *Rethinking Media Change: The Aesthetics of Transition,* edited by Henry Jenkins and David Thornburn, 39-60. Cambridge, MA: MIT Press, 2003.

Gunning, Tom. 'The World as Object Lesson: Cinema Audiences, Visual Culture and the St. Louis World's Fair, 1904.' *Film History* 6, no. 4 (1994): 422-444.

Hansen, Mark B.N. 'Media Theory.' *Theory, Culture & Society* 23, no. 2-3 (2006): 297-306.

Harmon, Leon. 'Analogs and Models of the Human Visual System.' *Optometry & Vision Science* 36 (June 1959): 304-312.

Harper's New Monthly Magazine. 'Editor's Scientific Record: Miscellaneous.' May 1879, 947.

Hauser, Ernst A. 'Raphael Eduard Liesegang 1869-1947' (Necrology). *Journal of Chemical Education* 26, no. 5 (May 1949): 274.

Hughes, Thomas. 'The Evolution of Large Technological Systems.' In *The Social Construction of Technological Systems: New Directions in the Sociology and History of Technology,* edited by Trevor Pinch, Thomas Hughes and Wiebe Bijker. Cambridge, MA: MIT Press, 1987.

Jameson, Dorothea and Leo Hurvich, ed., *Visual Psychophysics.* New York: Springer, 1972.

Johnson, V.E. *Modern Inventions.* New York: FA Stokes, 1915.

Kapp, Ernst. *Grundlinien einer Philosophie der Technik.* Braunschweig, Germany: Druck und Verlag von George Westermann, 1877.

Kern, Stephen. *The Culture of Time and Space, 1880-1918.* Cambridge, MA: Harvard University Press, 1983.

Lanterna Magica (Leipsig). 'Das Telelectroscop.' May 1898.

Liesegang, R.E., *Beiträge zum Problem des elektrichen Fernsehen.* Dusseldorf, 1891; 1899.

Lincoln Journal (Nebraska). 'Untitled' (Edison's Far-Sight Machine). May 26, 1889.

Lux, Fritz. *Der Elektrische Fernseher.* Ludwigshafen, 1903.

The Manufacturer and Builder. 'The Telectroscope Used for Photographing Over Telegraph Wires.' April 1879, 86.

Martin, Marcus. 'Television.' *Electricity.* October 21, 1921, 577.

Marvin, Carolyn. *When Old Technologies Were New.* Cambridge, UK: Oxford University Press, 1988.

Marx, Leo. *The Machine in the Garden: Technology and the Pastoral Ideal in America.* Cambridge, UK: Oxford University Press, 1964.

McLuhan, Marshall. *Understanding Media: The Extensions of Man.* Cambridge, MA: MIT Press, 1964.

Middleton, H. 'Seeing by telegraph,' letter to the editor, *Times* (London), April 24, 1880.

Mills, John. *Through Electrical Eyes: An Elementary Exposition of the Physics and Chemistry Involved in Television.* New York: AT&T/Bell Telephone Laboratories, March 1928.

Mitcham, Carl. *Thinking Through Technology: The Path Between Engineering and Philosophy.* Chicago, IL: University of Chicago Press, 1994.

Mitchell, WJT and Mark BN Hansen, eds. *Critical Terms for Media Studies.* Chicago, IL: University of Chicago Press, 2010.

Moffett, Cleveland. 'Seeing by Electricity.' *Pearson's Magazine,* October 1899, 490.

Le Mondes. 'Telectroscope.' January 16, 1879.

Morus, Iwan Rhys. "The Nervous System of Britain': Space, Time and The Electric Telegraph in the Victorian Age.' *The British Journal for the History of Science* 33, no. 4 (2000): 455-475.

Nason, C.H.W. 'A Short Course in Television.' *Television News,* March-April 1931, 30-31, 71.

Nelson, J.H. 'Ideal Seeing Conditions: The Study of the Human Visual System as a Basis for Prescribing Lighting.' *British Journal of Industrial Medicine* 2 (April 1945): 224.

Neue Weiner Tageblatt (Vienna, Austria). 'Der Elektrische Ferneher, besuch bei Hern Ein Jan Szczepanik.' March 17, 1898.

New York Times. 'Glasgow Listens to Sound of Faces.' February 4, 1927.

Olsson, Jan. 'Framing Silent Calls: Coming to Cinematographic Terms with Telephony.' In *Allegories of Communication: Intermedial Concerns from Cinema to the Digital,* edited by John Fullerton and Jan Olsson, 157-192. Bloomington: Indiana University Press, 2004.

Rasmussen, Nicolas. *Picture Control: The Electron Microscope and the Transformation of Biology in America, 1940-1960.* Stanford, CA: Stanford University Press, 1999.

Redmond, Denis D, 'An Electric Telescope,' letter to the editor, *English Mechanic,* February 7, 1879, 540.

Scientific American. 'Seeing by Electricity' (Carey). June 5, 1880, 355.

Scientific American. 'Siemens' Sensitive Electric Eye.' December 8, 1876: 374.

Scientific American Supplement. 'Prof. Rozing's 'Electric Eye' – A New Apparatus for Television.' June 1911.

Scientific American Supplement. 'The Telectroscope.' April 9, 1881, 4382.

Secor, H.W. 'Television, Of the Projection of Pictures Over a Wire.' *Electrical Experimenter.* August 1915, 131-2, 172-4.

Secor, H. W. 'Television Perfected at Last.' *Science and Invention,* June 1927, 108.

Senlecq, Constantin. 'The Telectroscope.' *The Electrician,* February 5, 1881, 141-2.

Shiers, George. *Early Television: A Bibliographic Guide.* London: Taylor & Francis, 1997.

Standage, Tom. *The Victorian Internet: The Remarkable Story of the Telegraph and the Nineteenth Century's Online Pioneers.* London: Weidenfeld & Nicolson, 1998.

Telegraphic Journal and Electrical Review. 'The Telectroscope.' February 15, 1879.

Times (London). 'The Telectroscope' (Senlecq). January 27, 1879 (reprinted in *Nature, The Electrician, Scientific American,* and *English Mechanic*).

T.R. letter to the editor, *English Mechanic,* July 31, 1891, 503-4.

T.R. 'Seeing by electricity,' letter to the editor, *English Mechanic,* June 26, 1891, 387-388.

Uricchio, William. 'Phantasia and Technè at the Fin-de-siècle.' *Intermediality: History and Theory of the Arts, Literature and Technologies* 6 (2005): 27-42.

Väliaho, Pasi. *Mapping the Moving Image: Gesture, Thought and Cinema Circa 1900.* Amsterdam: Amsterdam University Press, 2010.

Van der Vleuten, Erik. 'Large Technical Systems.' In *A Companion to the Philosophy of Technology,* edited by Jan Olsen, Stig Andur Pedersen, and Vincent F. Hendricks, 218-222. Malden, MA: John Wiley & Sons, 2012.

Van Loon, Joost. *Media Technology: Critical Perspectives.* New York: McGraw Hill, 2008.

Von Siemens, Werner. 'Action of Light on Selenium.' *Nature* 13, no. 334 (1876): 407.

Wade, Nicholas. *Visual Perception: An Introduction.* New York: Psychology Press, 2013.

Weekly Rocky Mountain News. 'Inventor of the Telectroscope.' May 12, 1898.

Willis, Martin. *Vision, Science and Literature, 1870-1920: Ocular Horizons.* New York: Routledge, 2015.

Zielinski, Siegfried. *Audiovisions: Cinema and Television as Entr'actes in History.* Amsterdam University Press, 1999.

Interlude

> In the future this dream will be realized, modern research assures us, but here on vastly different to anything yet made public. I differ here, for I believe the 'base principle' to be already within the present range of physics, but owing to the theoretical knowledge and apparatus required, it becomes more a matter for the physicist than the inventor. Whether this be so or not, who cares? None, save a few amateurs; and so it still remains like a good prize competition, open to all.
>
> -- E. August, 'Seeing by Electricity,' *English Mechanic*, 28 August 1891, 15.

> If we could only get one of the big research laboratories, like that of G.E.C. or the Western Electric Co. – one of those people who have large skilled staffs and any amount of money to engage on the business – I believe they would solve a thing like this [television] in six months and make a reasonable job of it.... For the ordinary amateur, however, it is not an easy class of experimental work, and would take a great deal of time, and probably cost a large amount of money.
>
> -- A. A. Campbell Swinton, 'The Possibilities of Television,' *Wireless World*, 23 April 1924, 118.

In the transition period between 1891 and 1924, the methods applied to solve the 'problem of television' changed radically even while the concept of television remained relatively stable. As August summarizes in his 1891 letter to the editor of English Mechanic, the preliminary legwork had been done. The nineteenth-century inventors and dreamers had accomplished everything within their means. But the machine still didn't work. The engineering of it would become a problem for physicists. For A. A. Campbell Swinton, an American physicist writing in 1924, the project was still so monumental that it was best left to corporations to tackle. G. E. or Western Electric, as far as Swinton was concerned, would bring the personnel and the resources to the table.

While the general concept of television remained the same as the nineteenth-century expectations of technology providing the ability to see over the horizon and thus annihilating space and its physical limitations, according to the rhetoric, the relationship between the user and the screen changed. Nineteenth-century depictions of seeing by electricity established the concept of television as a technology that could facilitate an intimate connection between loved ones at a distance. Popular visual culture presented television as a way to bring people closer together. As it passed into and back out of the physical laboratory, the way of describing the technology of television got in the way of these expectations in several ways. First, the rhetoric increasingly began to emphasize the electrical and electronic efficiency of energy involved in producing and reproducing a visual image on a screen and sending it to a distant place. Secondly, the quality of the picture on the screen grew more important than it had ever been in the past. As nineteenth-century documents on seeing by electricity rarely if ever mention the picture on the screen, expectations of picture quality grew from the aesthetics established by the cinema.

This transition from the nineteenth-century magic window to a Machine-Age functional electronic screen constitutes a transformation in the cultural meaning of technological mediation. As the satires of the telephonoscope and the 'Far-Sight Machine' testify, face-to-face interaction set the standard. The introduction of a screen-based technology threatened to destroy that relationship. The meaning of television as a way of seeing based on technological mediation emerged in the Machine Age when physicists made it possible to recognize the image of a person on an electronic screen.

When television made possible the annihilation of space, it turned out to be as much a perceptual change as a technological accomplishment. It would not be until much later that social psychologists recognized the gravity of this shift. A pivotal 1976 study, entitled the *Social Psychology of Telecommunications*, presented the theory of social presence that would redefine the meaning of face-to-face interaction for an age in which technological mediation was becoming the status quo.[1] Social-presence

1 J. A. Short, E. Williams, and B. Christie, *The Social Psychology of Telecommunications* (London: Wiley, 1976), 64-71; Frank Biocca et al., 'Toward a More Robust Theory and Measure of Social Presence: Review and Suggested Criteria,' *Presence* 12, no. 5 (2003): 456-480. *The Social Psychology of Telecommunications* (1976) introduced a concept of social presence as technologically mediated. It defines presence as a quality of the medium, establishing video as more socially present than telephony or writing. Research in screen interaction and electronically mediated communication constitutes an area known as computer-mediated communication, and integrates the study of social psychology. A sub-field has also emerged called presence research that extends the theory with a dual interest in psychology and technology. For a general introduction to computer-mediated communication scholarship and social psychology, see Eun-Ju Lee and Soo Youn Oh, 'Computer-Mediated Communication,' in *Oxford Bibliographies*, 15 January 2015; Nancy Baym, *Personal Connections in the Digital Age* (Malden, MA: Polity, 2010); Sherry Turkle, *Alone Together: Why We Expect More from Technology and Less from Each Other* (New York: Basic

theory upends the earlier way of thinking about face-to-face interaction as natural and replaces it with the power of a screen or other technological device to mediate between people at a distance. A medium like video, which establishes a visual and auditory bond between parties, supports more social presence than telephony, in which speakers can only hear each other's voices.

By the end of the nineteenth century, the prospect of seeing by electricity had ceased to be a popular topic in the press. While several stories trickled out in the first decade of the 1900s, the tone of these accounts reveals a subtle pessimism regarding the credibility and veracity of the new claims. Notably, the general interest in television persisted, as Swinton coined the phrase 'distant electric vision' in 1908, and a new generation of inventors, scientists, and engineers found pathways yet to be explored. During the time seeing by electricity turned into distant electric vision, it passed from the press and into the domain of the physical laboratory.

Television emerged fromf the laboratory into the public eye in 1920s America. Industry leaders including General Electric, RCA, and Bell Laboratories rolled out a hodgepodge of screens, cameras, and systems.[2] This brief moment in the long history of television was met with a mixture of anticipation and apprehension. The American public learned about television in popular-science magazines, in newspaper announcements, and in advertisements. A flurry of announcements declared the arrival of television, creating an atmosphere of anticipation for American audiences. At the same time, industry professionals also indicated their apprehension that television was not yet ready for broadcast.

The popular press rallied around a handful of inventors whose work represented the best efforts the industry giants could display. RCA promoted its own Vladimir Zworykin. AT&T had Herbert Ives at Bell Labs. GE had Ernst Alexanderson. Then there were several independents, such as Charles F. Jenkins, who ran W3XK, the first television station to broadcast in the

Books, 2011). Mark Poster and Katherine Hayles have extended the concept of electronic- and computer-mediated communication as it implies new forms of postmodern, post-human, or digital subjectivity. See Mark Poster, 'The Digital Subject and Cultural Theory,' in *What's the Matter with the Internet* (Minneapolis, MN: University of Minnesota Press, 2001); Katherine Hayles,'Virtual Bodies and Flickering Signifiers,' *October* 66 (1993): 69-91.

2 For more on the Machine-Age media landscape, see Abramson, *The History of Television, 1880-1941* (Jefferson, NC: McFarland & Company, 1987); Rasmussen, *Picture Control*; Barnouw, *Tube of Plenty*; Burns, *Television*; Gomery, *A History of Broadcasting in the United States*; Douglas, *Inventing American Broadcasting*; Herbert, *A History of Early Television*; Hilmes, *Only connect*; Kisseloff, *The Box: An Oral History of Television*; Magoun, *Television: The Life Story of a Technology.*

U.S., and John Logie Baird, an independent affiliated with the BBC. Each company advocated their own patented system, and a battle raged as to the form television would ultimately take.[3]

- Would consumers have television sets *in the home* or would television be *transmitted to public theatres*? (television for the home versus tele-cinema)
- Would television *broadcast* news and entertainment, or would it provide a *visual adjunct to the telephone*? (radio-vision versus two-way television)
- Would the *all-electronic* models win out, or would the *mechanical* system? (Sarnoff of RCA versus Ives of AT&T)

David Sarnoff, president of RCA, made a point, as many journalists and writers of popular science did, of distinguishing between different applications and systems of television. In an article written for *Modern Mechanics*, Sarnoff illustrated the different systems, suggesting that soon an 'ultimate system' would arrive (Figure 42).[4] The eventual success of the all-electronic models and the dominance of RCA in technological developments in television is an historical circumstance that, in retrospect, seems almost accidental. While historians of technology attribute the success of electronic television (and the corresponding demise of mechanical television) to a confluence of economic, industrial, and technological forces, I would add to that the cultural and discursive forces that worked to construct expectations about what television could and should do.

Sarnoff was the most outspoken and respected voice on the direction of the mass media. Confident advocate of the all-electronic system, Sarnoff believed that television would follow in the path of radio. He promoted a vision of the future in which television, radio, and cinema co-existed. His position as the famed 'General' should be recognized as much as a cultural trendsetter as an industrial leader, given the forcefulness of his vision for television alongside the power he wielded in establishing the direction for American mass communications.

Television historian Jeff Kisselloff conveys the story of how Sarnoff came to work at RCA. On his way to a job interview at the New York Herald, Sarnoff had taken the wrong turn off the elevator and ended up in the office of the Commercial Cable Company, where he was immediately offered a job as a courier. 'Years later, his son Robert Sarnoff would say, "Imagine what

3 'What Television Offers You,' *Popular Mechanics*, November 1928, 820-824.

4 David Sarnoff, 'Where Television Stands Today,' *Modern Mechanics*, April 1932, 40-46, 170.

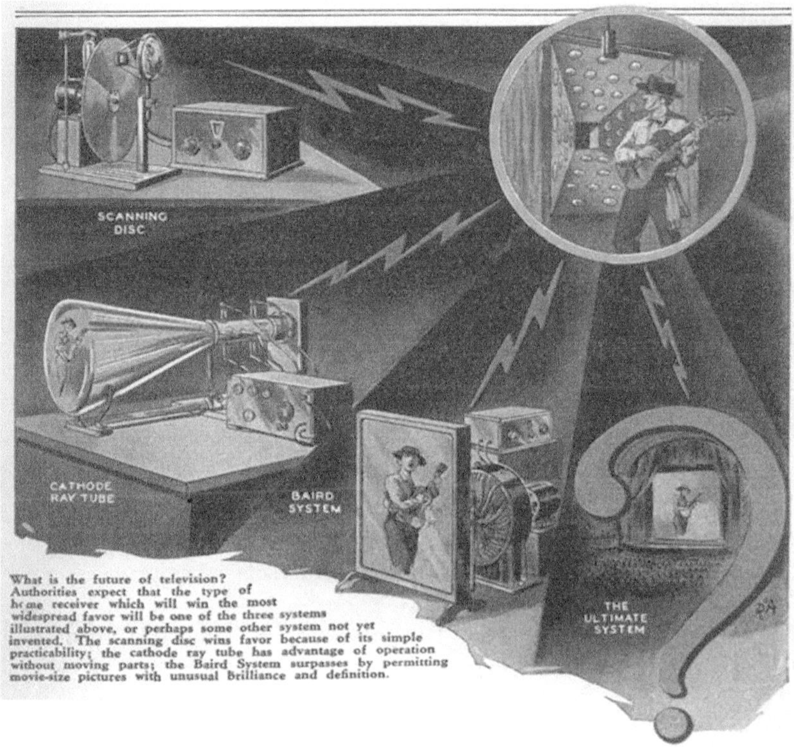

Figure 42. 'The Ultimate System,' in David Sarnoff, 'Where Television Stands Today,' *Modern Mechanics*, April 1932, 42.

would have happened had be turned left instead of right."'[5] It is this sense of contingency that I seek to identify in the story of Machine-Age television. In retrospect, the television industry grew up from the foundation of electronic systems and cathode ray tubes. But, in the context of the battles that raged in the 1920s, the outcome was anything but certain. What if Sarnoff had turned left instead of right?

RCA came out on top as the result of several industry decisions in radio and telephony, a careful balance of patents, monopoly, and the regulation of the airwaves. A key decision in 1930 dismantled the radio and telephone groups and designated which companies would have the right to develop television in the future. AT&T decommissioned its television project in order to enter into a lucrative deal with RCA, leasing their transcontinental lines. After 1930, the path forward for television was almost surely in the hands of

5 Kisseloff, *The Box*, 6.

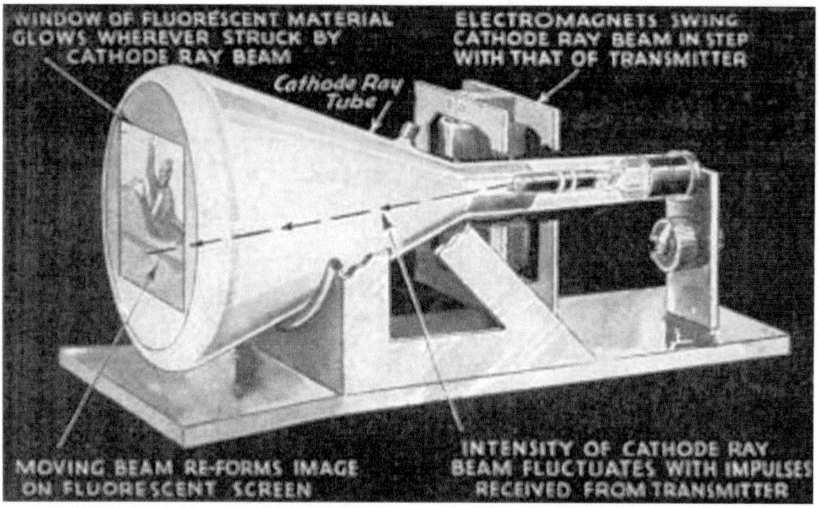

Figure 43. Zworykin's 'Iconoscope' Cathode Ray Tube (1933) in Alden Armagnac, '"Human-Eye" Camera Opens New Way to Television,' *Popular Science*, September 1933, 11.

RCA, David Sarnoff, and Zworykin, who advocated the all-electronic model that used cathode ray tubes (Figure 43).[6]

Bibliography

Abramson, Albert. *The History of Television, 1880-1941*. Jefferson, NC: McFarland & Company, 1987.

Barnouw, Erik. *Tube of Plenty: The Solution of American Television*. Cambridge, UK: Oxford University Press, 1990.

Baym, Nancy. *Personal Connections in the Digital Age*. Malden, MA: Polity, 2010.

Biocca, Frank, Chad Harms and Judee K. Burgoon. 'Toward a More Robust Theory and Measure of Social Presence: Review and Suggested Criteria.' *Presence* 12, no. 5 (2003): 456-480.

Burns, R.W. *Television: An International History of the Formative Years*. London: Institutions of Electrical Engineers, 1998.

Douglas, Susan J. *Inventing American Broadcasting, 1899-1922*. Baltimore, MD: Johns Hopkins University Press, 1989.

6 Kisseloff, 32; Abramson, *The History of Television*, 199; Magoun, *Television*, 57-59; Hugh Stollen, *Radio and Television Regulation: Broadcast Technology in the United States* (Baltimore, MD: Johns Hopkins University Press, 2000), 74-76. See also J. Earle Miller, 'What's Keeping Television Out of Your Home?' *Modern Mechanics*, May 1931, 98; Sarnoff, 'Where Television Stands Today.'

Gomery, Douglas. *A History of Broadcasting in the United States*. New York: Wiley-Blackwell, 2008;

Hayles, N. Katherine. 'Virtual Bodies and Flickering Signifiers.' *October* 66 (1993): 69-91.

Herbert, Stephen. *A History of Early Television*. New York: Taylor & Francis, 2004.

Hilmes, Michele. *Only Connect: A Cultural History of Broadcasting in the United States*. Boston, MA: Wadsworth Cengage Learning, 2013.

Kisseloff, Jeff. *The Box: An Oral History of Television, 1920-1961*. New York: Viking Press, 1995.

Lee, Eun-Ju and Soo Youn Oh. 'Computer-Mediated Communication.' In *Oxford Bibliographies in Communication*, January 15, 2015.

Magoun, Alexander. *Television: The Life Story of a Technology*. Westport, CT: Greenwood Publishing Group, 2007.

Miller, J. Earle. 'What's Keeping Television Out of Your Home?' *Modern Mechanics,* May 1931, 98.

Sarnoff, David. 'Where Television Stands Today.' *Modern Mechanics,* April 1932, 40-46, 170.

Popular Mechanics. 'What Television Offers You.' November 1928, 820-824.

Poster, Mark. 'The Digital Subject and Cultural Theory.' In *What's the Matter with the Internet.* Minneapolis: University of Minnesota Press, 2001.

Rasmussen, Nicolas. *Picture Control: The Electron Microscope and the Transformation of Biology in America, 1940-1960*. Stanford, CA: Stanford University Press, 1999.

Short, J. A., Williams, E., & Christie, B. *The Social Psychology of Telecommunications*. London: Wiley, 1976.

Stollen, Hugh. *Radio and Television Regulation: Broadcast Technology in the United States*. Baltimore, MD: Johns Hopkins University Press, 2000.

Turkle, Sherry. *Alone Together: Why We Expect More from Technology and Less from Each Other.* New York: Basic Books, 2011.

4. The Illuminating Engineers

Standardizing Vision

Abstract
In order to contextualize the theory and philosophy underlying Machine-Age mechanical television, Chapter Four explores the genesis of illuminating engineering in the early 20th century, its role in the efficiency movement, and the consequences it entails for the conception of the average viewer. Illuminating engineers adopted the Machine-Age philosophy of efficiency, struggling with standards and definitions that would establish a foundation for thinking about how the human, sometimes called a 'human seeing-machine', would adapt to life under electric light. Illuminating engineers constructed an image of the average observer, which served as the standard on which all models for interior electric lighting and television were measured. A pervasive attitude of control and efficiency governed and guided the further development of television technology.

Keywords: Standardization; efficiency; Machine Age; scientific management

A new image of television emerged during the Machine Age. Engineers devoted as much energy to designing systems and screens as they did to manufacturing a language and ideal image for the televisual viewer. Practitioners called themselves illuminating engineers. Coming from diverse backgrounds in physics, psychology, and design, these engineers worked towards the goal of making the televisual experience seem as natural as possible. In the process, they drew on established methods for colorimetry and photometry. Illuminating engineering practices relied on a conception of the human and the machine operating together in an efficient system. Illuminating engineers played as much a part in the design of screens and systems as they did in manufacturing an image of the ideal television viewer.

Roberts, I., *Visions of Electric Media: Television in the Victorian and Machine Ages.*
Taylor & Francis Group, 2019
DOI 10.5117/9789462986596_CH04

By framing the history of television as a technological transition, this chapter examines the work of engineers who contributed to making the televisual experience seem as natural as possible. In the process, these engineers manufactured a new way of seeing. By introducing the concept of the human-seeing machine, which relied on Machine-Age faith in efficient management and control, these engineers succeeded in manufacturing both the electronic screen and the ideal modern televisual viewing subject. This chapter looks at the culture, language, and philosophy of illuminating engineering through the work of three practitioners. Herbert Ives (1882-1953) took the physical stance. His colleague, Matthew Luckiesh (1883-1967), privileged the psychological view. Thirdly, Deane Judd's (1900-1972) work to standardize measurements for light and colour for the National Standards Bureau makes him a key contributor to the Machine-Age culture of illuminating engineering.

Ives, Luckiesh, and Judd represent a new generation of scientists and technicians who adopted the interdisciplinary practices of Illuminating engineering. As far as engineers were concerned, the photoelectric cell was the key component that made electronic television work. These 'electric eyes' transduced (or converted) light into electrical voltage, replacing the sluggish selenium cells. Think of it as the opposite of an electric light bulb. While the light bulb turns electricity into visible light, the photoelectric cell detects light and converts it into electrical current. For the layman, it was known as an electric eye, a kind of machine vision that did everything the human eye did, only better.

Along the way, a new language evolved to support television as a way of seeing, and vision as a kind of efficient work. It facilitated a way of thinking about television from two perspectives. In one sense, television was made up of hard, mechanical, and electrical properties and worked in a technical sense on principles like those of radio and telephony. But, from a viewer's perspective, it was also expressed as a way of seeing *through electrical eyes.* While scientists, engineers, journalists, and writers of popular science continued the progressive rhetoric of space annihilation, they also began to integrate a new way of thinking about television as a kind of technologically mediated vision.

The Machine-Age philosophy of efficiency drove this new rhetoric. What has been called a gospel, a cult, and a craze, the 'progressive era efficiency movement' found broad expressions across Machine-Age culture with the appearance of scientific management (Taylorism), as well as human and social engineering.[1] It also contributed to the formation of the National

1 On scientific management, see Samuel Haber, *Efficiency and Uplift: Scientific Management in the Progressive Era, 1890-1920* (Chicago, IL: University of Chicago Press, 1964); Kenneth

Bureau of Standards (NBS).[2] Politics and ideology motivated the broader efficiency movement, confronted by the perceived waste and corruption held over from the nineteenth-century Gilded Age.[3] As Jennifer Alexander puts it, these advancements had several things in common: 'Throughout the nineteenth and twentieth centuries, the notion of efficiency gained credibility by the breadth of applications to biology, economic thought, personal development, worker management, and social history.'[4] But a confluence of political, economic, and ideological forces at the beginning of the 20th century fostered efficiency as an ideal virtue. One classic study put it this way: 'the philosophy of efficiency is almost made to order for the progressive era mentality'.[5]

This atmosphere of progress, efficiency, and management that emerged in the Machine Age contributed to the formation of illuminating engineering. Before 1900, there was no consensus on the identity of the 'illuminating engineer'.[6] But in the Machine Age, it became a full-fledged profession driven by a new philosophy of efficiency. It contributed to the formation of a national society as well as to the standardization of practices and measurements. The new 'art' of illuminating engineering provided a philosophy that fuelled the scientific culture of television. This approach carries with it a new conception of the viewer, a new understanding of vision/seeing, and new implications for visual perception.

This chapter takes a media-oriented approach to the history of science and technology. Scholars who take this approach, such as Steve Wurtzler,

Thompson, introduction to *Scientific Management*, by Frederick Winslow Taylor (New York: Routledge, 2003); Sorin-George Toma, Ana-Maria Grigore, and Paul Marinescu, 'The Emergence of Scientific Management in America,' *Manager* 19 (2014): 127; Sharon Corwin, 'Picturing Efficiency: Precisionism, Scientific Management, and the Effacement of Labor,' *Representations* 84, no. 1 (2003): 139-165.

2 Rexmond Cochrane, *Measures for Progress: A History of the National Bureau of Standards*, (Washington, D. C.: National Bureau of Standards, US Department of Commerce, 1966).

3 Toma, Grigore, and Marinescu, 'The emergence of scientific management in America,' 128; Samuel P. Hays, *Conservation and the Gospel of Efficiency: The Progressive Conservation Movement, 1890–1920* (Cambridge, MA: Harvard University Press, 1959), Chapter 7; Jennifer Alexander, *Mantra of Efficency: From Waterwheel to Social Control* (Baltimore, MD: Johns Hopkins University Press, 2008), 5.

Jennifer Alexander, 'Efficiencies of Balance Technical Efficiency, Popular Efficiency, and Arbitrary Standards in the Late Progressive Era USA,' *Social Studies of Science* 38, no. 3 (2008): 325.

4 Jennifer Alexander, *Mantra of Efficiency*, 3.

5 Haber, ix.

6 Otter, *The Victorian Eye.* Otter shows how illuminating engineering existed before 1900, in a variety of practices and methods. Universal standards for measurement and practice did not coalesce until after the formation of its professional societies, around 1910.

Jonathan Sterne, and Mara Mills, for example, integrate research in science and technology into an examination of media history.[7]

Embracing the broad range of research and scholarship in the history of media and technology, this micro-history also takes advantage of primary sources, including scientific journals, popular-science magazines, educational pamphlets, and monographs.[8]

The history of illuminating engineering informs the study of television. But thinking about the role that the design and manufacture of artificial lighting played in the history of the media also bears on our knowledge of video and audiovisual compression in a broader sense. By looking at how these engineers defined average brightness and balanced the levels of electric transmission against the goal of reproducing an adequate picture quality is also connected to our understanding of contemporary compression practices. For example, one could ask how the goals of such a balance have changed over the past hundred years. Illuminating engineering practices informed the early history of television and, as such, provide a foundation for the study of video-compression and image-processing standards today. The MPEG format group, for example, established the protocols for image and video compression in the late 1990s based on the standards established in the 1930s and the philosophy of efficiency and manipulating faults in human visual perception, all of which I show to have been established by 1930.[9]

7 Steve Wurtzler, *Electric Sounds: Technological Change and the Rise of Corporate Mass Media* (New York: Columbia University Press, 2007); Jonathan Sterne, *MP3: The Meaning of a Format* (Durham, NC: Duke University Press, 2012); Mara Mills, 'Deafening: Noise and the Engineering of Communication in the Telephone System,' *Grey Room* 43 (2011): 118-143. See also James Beniger, *The Control Revolution: Technological and Economic Origins of the Information Society* (Cambridge, MA: Harvard University Press, 1986); David Mindell, *Between Human and Machine*; Lewis Mumford, *Technics and Civilization* (New York: Harcourt, 1934); John Jordan, *Machine-Age Ideology: Social Engineering and American Liberalism, 1911-1939* (Chapel Hill, NC: University of North Carolina Press, 1994); Otter, *The Victorian Eye*; David Nye, *Electrifying America: Social Meanings of a New Technology, 1880-1940* (Cambridge, MA: MIT Press, 1992); Sean Johnston, *A History of Light and Colour Measurement: Science in the Shadows* (Bristol, UK: Institute of Physics Publishing, 2001); Sean Cubitt, *The Practice of Light: A Genealogy of Visual Technologies from Prints to Pixels* (Cambridge, MA: MIT Press, 2014); Abramson, *The History of Television*; R. W. Burns, 'Prophecy into Practice: The Early Rise of Videotelephony,' *Engineering Science and Education Journal* 4, no. 6 (1995): S33-S40.

8 Science and other technical periodicals and proceedings include *Bell Systems Technical Journal (BSTJ)*, *IEEE*, *Transactions of the Illuminating Engineering Society*, *Journal of the Optical Society of America (JOSA)*, and *Nature*.

9 Jonathan Sterne and Dylan Mulvin, 'The Low Acuity for Blue: Perceptual Technics and American Color Television,' *Journal of Visual Culture* 13, no. 2 (2014): 118-138; Barbara Saunders and Jaap Van Brakel, 'The Trajectory of Color,' *Perspectives on science* 10, no. 3 (2002): 302-355; Janos Schanda, ed. *Colorimetry: Understanding the CIE System* (New York: John Wiley & Sons,

While there are several similarities to be noted between the prehistory of video processing and Jonathan Sterne's work on the history of MP3 encoding, my investigation looks further back to a time before the concept of 'video'.[10] The term itself cannot be found in print before the mid 1930s. Without a concrete term for the process of moving-image transmission, illuminating engineers described it as a technical and scientific procedure in which light converted into electricity resulted in a perceptible image.

Illuminating Engineering

The practices of illuminating engineering date back to the nineteenth century, when scientists, technicians, and social reformers sought to organize a better way of lighting public spaces.[11] It focussed primarily on the efforts of city planners in designing public architecture according to the capabilities of gas lighting, on the one hand, and representatives of gas companies in measuring as well as advising their customers in safety and best practices. But, at the dawn of the 20th century, illuminating engineering identified a new set of goals. The efficient production of light became their newfound motivation. This new goal facilitated a common bond among practitioners and helped their ranks develop into a professional society. The Illuminating Engineering Society of America was founded in 1906, published their first volume of transactions that year, and held their first annual convention in 1907. For all intents and purposes, they claimed, illuminating engineering simply did not exist before 1900. According to this view, the profession required a concrete and universal handbook of practices, which assured continuity, uniformity, and standards. The formation of a professional society and its journal coincided with the appearance of college courses on the subject. All of these events coalesced to make illuminating engineering a hot topic.

The illuminating engineers congregated around several nodes. In America, the Optical Society (OSA), the National Electric Light Association (NELA), the United Gas Association, and the National Bureau of Standards (NBS) funded laboratory research and supported a professional community of illuminating engineers. Unlike other sciences, illuminating engineering

2007); Stefan Winkler, *Digital Video Quality: Vision Models and Metrics* (New York: John Wiley & Sons, 2005); Hong Ren Wu and Kamisetty Ramamohan Rao, eds. *Digital Video Image Quality and Perceptual Coding* (Boca Raton, FL: CRC press, 2005).

10 The OED and other reference dictionaries cite the word 'video' originating in English-language print in the 1930s.

11 Otter, 2008; Johnston, 2001.

was a distinctly interdisciplinary endeavour, requiring the collaboration of many fields, principally physics (optics), physiology (vision), and psychology. The first annual conference of illuminating engineering in 1911 marked the beginning of a push towards the standardization of light and vision.

The Illuminating Engineering Society preached a philosophy of efficiency similar to other management and organizational initiatives of the time. Consistent motifs provided the backbone for efficiency applied to situations across different levels of society, including the urge to quantify and standardize. Commonalities such as this serve to illustrate how the ideological impetus established methods of efficiency that impose 'arbitrary' standards on otherwise natural, physical laws or limits.[12] The 'gospel of efficiency' provided the foundation for a fusion of scientific and industrial forces to shape the 'work' of 'man'.[13] The notion of the ideal human and the efficient worker that this mindset fostered in popular culture aligned with a thermodynamic model. It treated the human body like a machine. The notion that the body becomes fatigued provided a link between the well-oiled machine and the managed labourer. As a metaphor, it supported a conception of the body as a machine and the work as a practice to be engineered, managed, and standardized. A utopian image of scientific, technological, and industrial progress infused the efficiency movement and illuminating engineering, articulated in the theme of Chicago's 1933 Century of Progress World's Fair: 'science finds, industry applies, man conforms'.[14]

Several scholars who study this moment in American history emphasize how the political, scientific, and industrial movements set standards unrelated to physical laws. Jennifer Alexander calls the new standards set in place by the many branches of the efficiency movement 'arbitrary', referring to the social and cultural construction of limits of work, efficiency, and fatigue unrelated to natural, physiological, or physical thresholds.[15] Similarly, historian of technology Sean Johnston has noted how standards of light and

12 Alexander, 6, 13, 78.

13 Alexander, 'Efficiencies of Balance,' 325.

14 *Official Guide: Book of the Fair* (Chicago, IL: Century of Progress International Exposition, 1933). For a discussion of the motto, see Robert Rydell, *World of Fairs: The Century-of-Progress Expositions* (Chicago, IL: University of Chicago Press, 1993).

15 Alexander, *Mantra of Efficiency*; Lennard J. Davis, 'Constructing Normalcy: The Bell Curve, the Novel, and the Invention of the Disabled Body in the Nineteenth Century,' in *The Disability Studies Reader*, ed. Lennard J. Davis (New York: Taylor & Francis, 2006), 3-16; Waltraud Ernst, 'The Normal and the Abnormal: Reflections on Norms and Normativity,' in *Histories of the Normal and the Abnormal: Social and Cultural Histories of Norms and Normativity*, ed. Waltraud Ernst (New York: Routledge, 2006), 1-25; Kurt Danziger, *Constructing the Subject: Historical Origins of Psychological Research* (Cambridge, UK: Cambridge University Press, 1994).

colour were established during this time in response to what was perceived as a new problem in scientific and technical circles, 'a problem substantially created and solved in the interwar period'.[16] In a very short amount of time, problems such as poor lighting and industrial waste were assessed based on the vision of an ideal work place and the efficient worker. The standards set in place during this time became unquestioned tenets that continue to provide guidance today in many areas of management. In colour management and lighting standards, for example, the standards established during the interwar period continue to provide the foundation for colour and brightness today.[17] The dramatic difference between natural and culturally constructed standards of light and colour calls into question the way human perception adapts to such standards. Over time, such artificial limitations begin to seem natural. They become unquestioned aspects of the human world.

Along with the efficiency movement, a rift between the practitioners of physics and psychology in the scientific community provides an important context for the standards that were set in these fields. They were simply at odds in defining the differences between the physical action of light and the perception of brightness. Particularly in the new field of illuminating engineering, conflict between the physical and psychological views came across most clearly in the language employed. Whereas a physicist would discuss the reflections of light radiation, a psychologist would emphasize the perception of brightness and colour. The illuminating engineers rallied to bring these communities together, in effect constructing a new language of light, colour, and vision that allowed these conflicting views to communicate with one another. The dominance of the physical view at the time led to a preference for the measurement of light quantities and objective colour measurements. In turn, they contributed to the quantification of a standard observer and colour space. These standards provided the foundation for developments in bandwidth restrictions and established methods for image processing and audiovisual signal processing.[18]

Work at NELA

In the first decades of the 20th century, relatively few practitioners identified themselves specifically with the field of Illuminating engineering. Their

16 Johnston, 160.

17 See Saunders; Johnston.

18 See Johnston, 159-160.

philosophy and language developed most clearly in professional gatherings as well as in the research that went on in the few industrial laboratories that existed. Apart from the proceedings of the IES, illuminating engineers performed their work and documented their progress at a small number of sites across the country. The National Electric Lamp Association (NELA), a research division of GE, was one such site. In Cleveland, Ohio, NELA's physical laboratory employed a handful of illuminating engineers, including Herbert Ives and Matthew Luckiesh.

While they claimed to practise the same method, their approaches could not have been more different.[19] Ives preferred the physical view.[20] In an article entitled 'An Illuminating Engineer's Conception of An Ideal Light', he identifies 'the distinct function of the illuminating engineer' as 'the utilization of light sources in the most efficient manner'. Ives, along with others, emphasized the interdisciplinarity and collaboration involved in practising illuminating engineering:

> He finds it necessary to introduce a number of factors—physiological, psychological, aesthetic—which vastly complicate the process of arriving at high efficiency, or of expressing that efficiency in definite terms. It is in fact the addition of these non-physical factors that makes illuminating engineering a science or art, or artistic science, apart from the science of light production.[21]

19 Leonard Keene Hirshberg, 'Why Our Eyes Very Often Deceive Us,' Newspaper clipping, *Evening Sun*, 20 July 1914, Container 4, Frederic Eugene Ives and Herbert Eugene Ives Papers, Manuscript Division, Library of Congress, Washington, D. C.; 'Physical Laboratory of the National Electric Lamp Association: A Laboratory for Research in Illumination and Electric Lighting,' *Electrical Review and Western electrician*, 10 September 1910, 507-511; Induction Letter, Illuminating Engineering Society, 25 September 1908, Container 9, Frederic Eugene Ives and Herbert Eugene Ives Papers, Manuscript Division, Library of Congress, Washington, D. C.; Oliver Buckley and Karl K. Darrow, 'Herbert Eugene Ives (1882-1953),' Washington, D. C.: National Academy of Sciences, 1956; See also Charles Rubinstein, 'Optics at Bell Laboratories – General Optics, television, and Vision,' *Applied Optics* 11, no. 11 (1972): 2401-2411.

20 Ives' contributed to photometry and colorimetry in the first two decades of the 20th century, setting the stage for the standardization of light and colour vision in the 1930s. Several of his studies and measurements proved essential to the formulation of what is known as the 'standard observer', a quantified measurement of 'average' brightness and colour vision. The titles of his many lectures testify to his prejudices towards the physical properties of light and colour measurement, as opposed to the approached taken by psychologists and physiologists who privileged the perceptual role of vision: 'Vision as a Physical Process,' 'The Establishment of Photometry on a Physical Basis,' 'Evidence that Persistence of Vision is a physical conduction Phenomenon.'

21 Ives, 'An Illuminating Engineer's Conception of an Ideal Light,' 419.

While illuminating engineers sought to foster a community of collaborators, in practice, those efforts fell short. Ives described light as a physical phenomenon while Luckiesh preferred to think of it as the efficient work of seeing.

Ives based his conception of ideal light firmly in the principles of efficiency, establishing the agenda for standards of lighting quality on the assumption that interior lighting should simulate natural conditions under sunlight: 'an excellent argument can be made for daylight as the ideal illuminant on the ground that it is the light on which the human race has been reared'.[22] While aesthetics played a part, for Ives, the goal was always efficiency of energy and light.[23] In comparison, his colleague Matthew Luckiesh advocated for a conception of illuminating engineering as a method for making seeing easier and more efficient. Recognizing that electric light was far from a natural environment for the human eye, he thought of illuminating engineering as a way to mitigate the artificiality of seeing by electric light. Luckiesh built on Ives' notion of the standard of natural daylight, but pushed the thesis further into the realm of simulating brightness. As a result, Luckiesh's version of illuminating engineering emphasized seeing as a practice or science distinct from the efficient production of light. Luckiesh staked his reputation on advocating for more light. He believed that, in order for humans to adapt to artificial interior environments, they needed an amount of light comparable to natural daylight.

Luckiesh referred to the observer as a 'human seeing-machine'.[24] His particular approach differentiated between seeing, a cognitive ability, and vision, a fusion of the physiology of the eyes and electrical lighting. He defined the human subject as a mechanical 'device' designed to fit into its environment: 'The human being is a human *device* which does the seeing.'[25] For the utilitarian engineer, eyes were like the hands that did the labour. Just as hands can be trained to pick up rhythms and motions that become second nature, so too can the eyes adjust to artificial lighting conditions.[26] Seeing, like other kinds of labour, could be controlled and optimized using

22 Ives, 'Ideal Light,' 424.

23 He emphasizes this in his conclusion, 429, which is where his thesis becomes distinct from Luckiesh's interpretation.

24 Matthew Luckiesh and Frank Moss, 'The Human Seeing-Machine,' *Journal of the Franklin Institute* 215 (June 1933): 629-654; Matthew Luckiesh and Frank Moss, 'The New Science of Seeing,' *Transactions of the Illuminating Engineering Society* (January 1930): 15-39; Matthew Luckiesh and Frank Moss, *The Science of Seeing* (New York: Van Nostrand Co., 1937).

25 Matthew Luckiesh and Frank Moss, *The New Science of Lighting* (Lighting Research Lab, General Electric, 1934), 12

26 Cady, 247; Nye, 362-364.

the principles of engineering. Luckiesh integrated this ideology into his definition of 'seeing':

> The science of vision is concerned with the abilities and limitations of eyesight as a tool. Seeing involves this tool and another tool—lighting. But seeing is much more than this. It is an activity of human beings operating as human seeing-machines. Seeing is work that a human being does. It requires lighting and vision but also human effort.[27]

Ultimately, Luckiesh wanted to design a better human being: a human seeing-machine optimized for life under electric light.

Presenting at an MIT congress of technology and industrial efficiency in 1911, illuminating engineer John Codman characterized the new language that was emerging in his trade:

> To a considerable extent, a new terminology has been evolved and technical names used only by the scientists five or six years ago [1905-6], may now be seen in common use in the magazines devoted to gas, electricity and illumination and may be heard in the mouths of commercial men.[28]

Illuminating engineers faced the challenge of negotiating between two different languages. They were cast as interdisciplinary scientists, technicians who traversed the worlds of psychology (vision) and physics (light). One handbook described it as a praxis that combined art and science, design, and engineering:

> The illuminating engineer is in a somewhat anomalous position. As an engineer he has to deal with engineering materials and sources of energy, using them to obtain results in the most efficient and economical manner; but like the artist, he has in specifying his requirements and in judging his results to satisfy that most capricious organ, the human eye.[29]

27 Matthew Luckiesh and Frank Moss, 'The New Science of Seeing,' *Transactions of the IES* (January 1930), 3-4. The Illuminating Engineering Society (IES) ranked this paper in the 100 most significant papers of the 20th century: 'The 14 articles in this category [Vision] describe various important effects and applications of vision research on lighting practice between 1910 and 1999.' '100 Significant Papers,' *Illuminating Engineering Society*, ies.org.

28 John Codman, 'Advent of Illuminating Engineering,' in *Technology and Industrial Efficiency* (New York: McGraw-Hill, 1911): 278; 'A Year's Progress in Illuminating Engineering,' *Good Lighting and Illuminating Engineer* 7 (1912): 2.

29 Leonard Jolley, John Waldram, and George Wilson, *The Theory and Design of Illuminating Engineering Equipment* (New York: Wiley, 1931), 3.

Their practical approach privileged a common-sense attitude to explaining lighting design and visual aesthetics. While optics, colour science, and physiology provided essential foundations, illuminating engineers were taught to move past the hard science in order to find a happy medium between technical and vernacular languages, between applied physics and aesthetic design.[30]

Within the 1920s scientific community, the physicalist view dominated the psychological. As historian of science Sean Johnston describes, the schism between physics and psychology ran deep.[31] Physicists spoke in the language of objectivity. Psychologists preferred to speak of light in terms of perception, a way of thinking about vision that more closely aligned with the vernacular. The two even defined their object of study differently. Physicists studied 'spectral luminosity'; psychologists referred to 'visibility'[32]

The National Bureau of Standards

Complicating matters further, light itself is invisible; human eyes can only see the objects off which light bounces. Since vision cannot function without light, the colour of objects necessarily comes into play when attempting to measure brightness. Illuminating engineer Deanne Judd (1900-1972), a representative for the National Bureau of Standards, claimed that *hetero-chromatic spectrometry*, measuring brightness by the colour of light, was the greatest challenge facing his science. To illustrate the challenge, he told the story of Jack and Jill, a married couple arguing over which brand of spread to serve at the dinner table.[33] Jack prefers the taste of margarine because he got used to it during the war. Jill prefers butter. But Jill won't

30 See, for example, the textbooks and course titles from the early illuminating engineering college courses. It was standard to include introductory chapters on optics, physiology, and physics, and a second part on practice and design. Francis Cady, *Illuminating Engineering* (New York: Wiley, 1925); Leonard Jolley, John Malyon Waldram, and George Humphreys Wilson, *The Theory and Design of Illuminating Engineering Equipment* (New York: Wiley, 1931); *Light: Its Use and Misuse* (New York: Illuminating Engineering Society, 1912); *IES Lighting Handbook: A Standard Lighting Guide* (New York: Illuminating Engineering Society, 1947); 'The Johns Hopkins University Course of Lectures on Illuminating Engineering,' *The Illuminating Engineer* 5 (1911): 256-257.

31 Sean Johnston, *A History of Light and Colour Measurement*, 158-159; Sean Cubitt, *The Practice of Light*, 139.

32 Lloyd Jones, 'The Historical Background and Evolution of the Colorimetry Report of the OSA,' *Optometry and Vision Science* 21, no. 8 (1944): 325.

33 Deane Judd and Günter Wyszecki, *Color in Business, Science and Industry* (New York: John Wiley and Sons, 1952), 343.

compromise with Jack because of the margarine's unappetizing colour: an ugly white. They quarrel over some alternatives, including changing the colour of the tablecloth to offset the colour of the food, and even changing the colour of the light bulbs. Judd tells the story to show how colouring the margarine would provide the easiest solution to the problem. Industry can control consumer perception by manipulating the colour of their products.

The story demonstrates the difference between the objective colour of an object and the viewer's perception of it. As Judd explains, 'In psychophysical terms dominant wavelength, purity, and reflectance refer to the light reflected by a specimen. White and yellow are psychological terms describing the color perceived to belong to the specimen.'[34] In this view, the objective colour of the product, as judged by the scientific instruments, matters less than the aesthetic image perceived by the consumer. Both Judd and Luckiesh spoke of illuminating engineering as a method for controlling visual perception. Judd spoke of manipulating the colour and brightness of goods to affect the consumer's consciousness of products. Luckiesh described the subject as a 'human seeing-machine' and the goal of the illuminating engineer to design the artificial interior environment to make the work of seeing as effortless as possible. As illustrated in Figure 44, 'seeing is a partnership between vision and lighting'. For illuminating engineers, the practice of designing perfect lighting environments came down to the matter of control. While lighting specialists have complete control over the environment given their mastery of the illuminating engineering practices and opticians have complete control over the mediating devices used to see, the process of vision itself falls outside the realm of control. Luckiesh's understanding of seeing as 'a partnership of vision and lighting' demonstrates the struggle illuminating engineers faced in negotiating between the subjective realm of the mind and the objective world of physics.

The Standard Observer

It took an international community engaged in the measurement of light and colour to compile the research necessary to quantify the sensitivity of the human eye. But that would not have been enough to solve the problem of television. The final stage of the process, engineered by Machine-Age scientists according to the particular philosophy of efficiency and progress, cast the human viewer in the role of a 'standard observer'.

34 Judd, *Color in Industry*, 342.

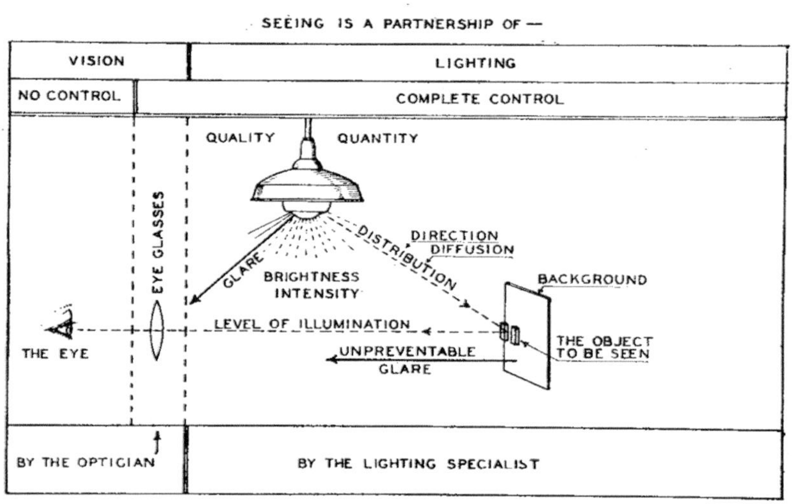

Figure 44. 'Seeing is a partnership of vision and lighting,' in Matthew Luckiesh and Frank Moss, 'The new science of seeing,' *Transactions of the Illuminating Engineering Society*, January 1930, 17.

Based on studies of the eye's sensitivity to light and colour, the International Commission on Illumination (CIE, *Commission Internationale de l'Eclairage*), an organization tasked with setting standards for lighting, developed what is known as the 'standard observer'. This international group of physicists, psychologists, and physiologists was tasked with adopting standards for light and colour to be used in science and industry. A definitive 1931 decision, preceded by a 1922 preliminary study, has remained the authority on standards of light and colour ever since.[35] The 1931 standard was synthesized from the data presented in just two studies, together consisting of just seventeen male subjects.[36]

35 Deane Judd, 'The 1931 I. C. E,. Standard Observer and Coordinate System for Colorimetry,' OSA 23, no. 10 (1933): 359-373; Deane Judd, 'Fundamental Studies of Color Vision from 1860-1960,' *NAS Symposium* 55 (1966): 1313-1330; Lloyd Jones, 'The historical background and evolution of the colorimetry report of the OSA,' *Optometry and Vision Science* 21, no. 8 (1944): 317-333; Janos Schanda, *Colorimetry* (New York: Wiley, 2007); Sean Johnston, 'The Construction of Colorimetry by Committee,' *Science in Context* 9 (1996): 387-420; Michael Brill, 'How the CIE 1931 color-matching functions were derived from Wright-Guild data,' *Color Research & Application* 23, no. 4 (1998): 259-259.

36 Johnston, 1996, discusses data sets: American investigators had made determinations of the 'standard observer' in 1912, 1917, and 1923 on progressively larger samples of people. The last NBS results, on 52 individuals aged under 30, measured in 'good lighting conditions', were proposed to the CIE as the response of a 'standard observer'.' 'The experimental data used in the 1931 color system were based on the observations of only 17 British subjects.'

The standard observer is not a person like a television viewer or even a human seeing-machine. It is a mathematical model constructed by a committee of scientists. In the literature, the 'standard observer' is represented in charts and diagrams like those in Figure 45. Johnston elaborates:

> Only the highly artificial 'standard observer'—a table of numbers representing the response of a typical eye to the three reference colours—related this physical approach to visual perception. The acrimony in the subject through the remainder of the decade related to this restrictive physical definition of the subject.[37]

The standard observer provided the model for illuminating engineers, physicists, electrical engineers, and their colleagues at the national board of standards to assign and enforce judgments about the limits of human perception. This mathematical model justified setting standards for the brightness of light and the range of visible colour that would be reproduced and reinforced by industry and science.

As Alexander, Johnston, and others attest, the ideology promoted by Machine-Age science and industry reconfigured the balance between nature and culture.[38] The construction of the 'standard observer', as illustrated in several works by Judd, the NBS-appointed representative to the CIE, illustrates what Alexander meant by 'arbitrary standards': a political ideology of control applied a culturally constructed and technologically facilitated way of seeing to engineering. The 'standard observer' represents a culturally and historically specific construction of vision, founded on 'arbitrary' standards of brightness and the quantification of the colour spectrum.

The challenge that light and colour measurement poses to the division between nature and culture calls into question the basic notions of normal, average, and natural. Over the course of almost a century, this standardized way of seeing has come to seem natural. Its mechanisms are engineered into the everyday devices and tools in ways designed to go unnoticed: the automatic dimming of a computer screen, the uniform brightness of windowless office buildings. Designed according to universal standards, the modern artificial environment in which we live became naturalized over a long process of accommodation.

37 Johnston, 1996, 173; Richard Hunter and Richard W. Harold, *The Measurement of Appearance* (New York: John Wiley & Sons, 1987), 81.

38 Alexander 2008; Johnston 2001; Saunders 2002; Cubitt 2014.

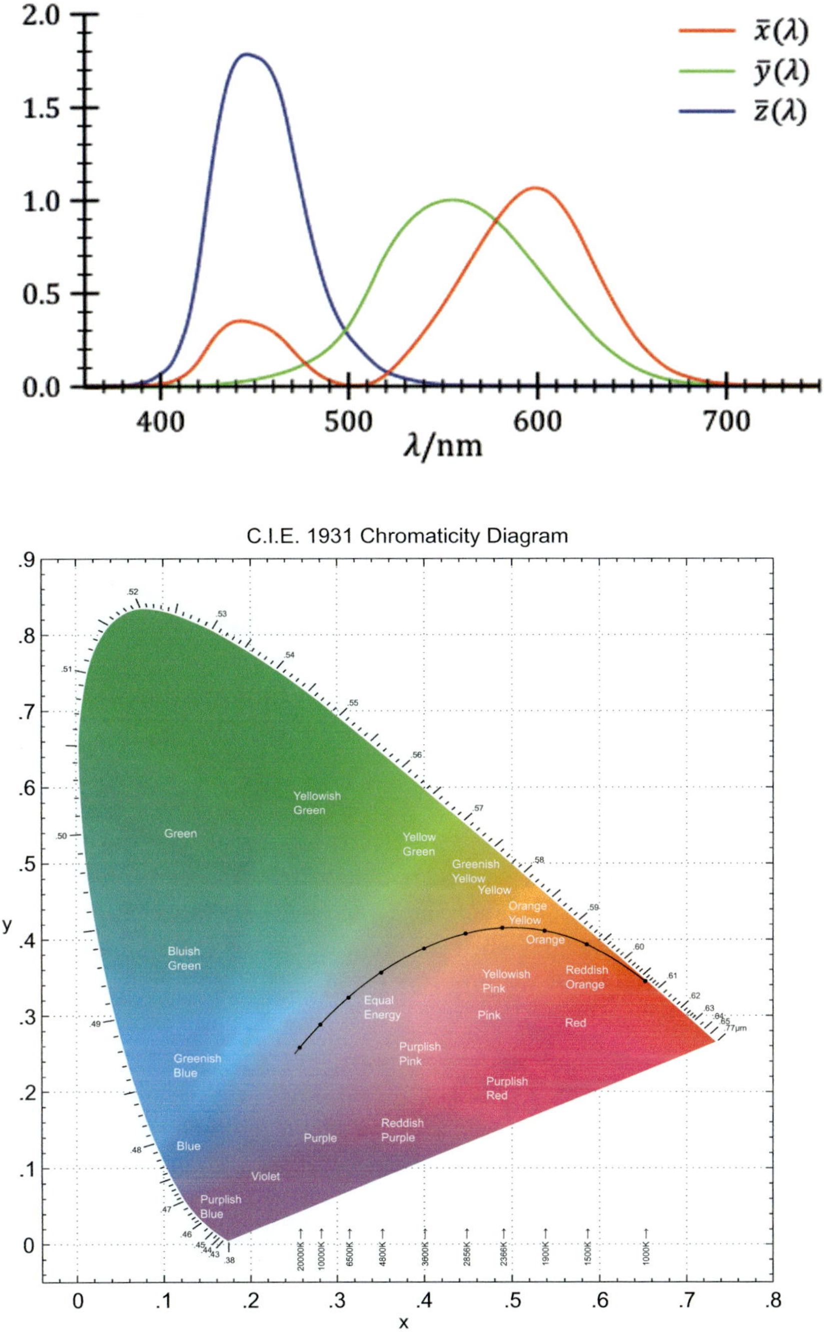

Figure 45. Representation of the CIE standard observer, colour-matching functions, CIE standard colour space.

Cultures manufacture 'normal' just like technology, science, and industry manufacture what passes for 'natural'. Through a process of accommodation, the artificial comes to seem commonplace. The notion of engineering the human went hand in hand with designing the modern, artificial world.

Historian of science Waltraud Ernst describes this process of accommodation and artificiality as a conflict between what naturally 'is' and what scientists and engineers think it 'ought to be'.[39] The *is* and the *ought* refer to two different kinds of normal: the first, a transparent description of a norm, and the second, a prescription. While some social norms arise organically, institutions form and shape social groups through active processes. *Social norms*, then, contrast with *quantitative standards*: artificial constructions.

Electric Eyes

Luckiesh's termed 'human seeing-machine' is a rough characterization of the Machine-Age understanding of seeing as work. The idea that the human was a seeing-machine also underpinned the discussion of television and its human viewer. Nowhere was that analogy more apparent than in the engineering of the photocell, known in popular culture as the electric eye. A feat of modern science, electric eyes harnessed invisible rays and turned them into exploitable energy:

> For not only have we harnessed the electron, but also, with the aid of the photo-electric cell, we have learned how to build light beams into our modern machines, in much the same way as we would use steel, aluminum, or glass. Light, when you come to think of it, is an idea mechanical material for many purposes, being weightless, substanceless, frictionless, and instantaneous.[40]

Technically, the photocell is a kind of transducer. It turns light into electric current. What the microphone does for sound, the photocell does for light.

The electric eye marks the threshold between television's *is* and its electronic *ought*. As one journalist noted:

> Immortal fame and presumably a vast fortune awaits the fortunate individual who can take *television as it is and make it what it ought to be.*

39 Waltraud Ernst, 'The Normal and the Abnormal,' 3-4; See also, Danziger, *Constructing the Subject.*
40 O. H. Caldwell, 'The Tiny Electron Works Wonders for Man,' *New York Times,* 6 July 1930.

> Someday, perhaps in the not so distant future, we may have a television receiver that will do for your eyes what the broadcast receiver now does so well for your ears.[41]

Electronic systems elicited reactions ranging from incredulity and scepticism, to awe and congratulation. The mere ability to transmit moving pictures, however, did not live up to the expectations of image quality established by the cinema. Viewers wanted a realistic image that lived up to the pictures their imaginations produced to accompany radio dramas. According to most, television had a long way to go before it reached what it 'ought to be'. The difference between the passive transmission of light and the simulation of distant visions marks the threshold between is and ought.

An example of an is/ought distinction can be found in comparing the notion of a threshold with that of the optimum. Nineteenth-century scientist Gustav Fechner studied the thresholds of vision, lending his name to the law of visual accommodation to brightness (Figure 46). Fechner's law provides a basic principle for modern optical science. The principle reinforces visual perception as a subjective measurement, complicated by the fact that human vision adapts to the brightness of its surroundings. It shows that visual adaptation functions relative to the environment. One psychophysics textbook on measuring 'appearance' describes the visual threshold like this:

> In the understanding of vision, the most important application of this principle is in the way we see lightness. For example, if we look at three ceramic tiles in a row, a white one reflecting 80% of the incident light, a light gray one reflecting 40%, and a dark gray one reflecting 20%, the eye sees lightness *difference* between adjacent tiles as about equal. This is because the change from 80 to 40% is 50%, and the change from 40 to 20% is about 50%. On a log-reflectance scale these equal ratios become equal increments. The name Fechner is associated with this fact. When expressed on the log basis the term Weber-Fechner law is often used.[42]

Figure 47 provides a visual depiction of this difference. In layman's terms, the principle relates to the perception of brightness. As brightness increases, perception of it slowly levels off. The eye's natural adaptation to brightness can be measured on Fechner's curve. 'Is' merely describes the phenomenon.

41 Alfred Lane, 'The Real Facts About Television,' *Popular Science,* September 1928, 43.
42 Hunter and Harold, 46.

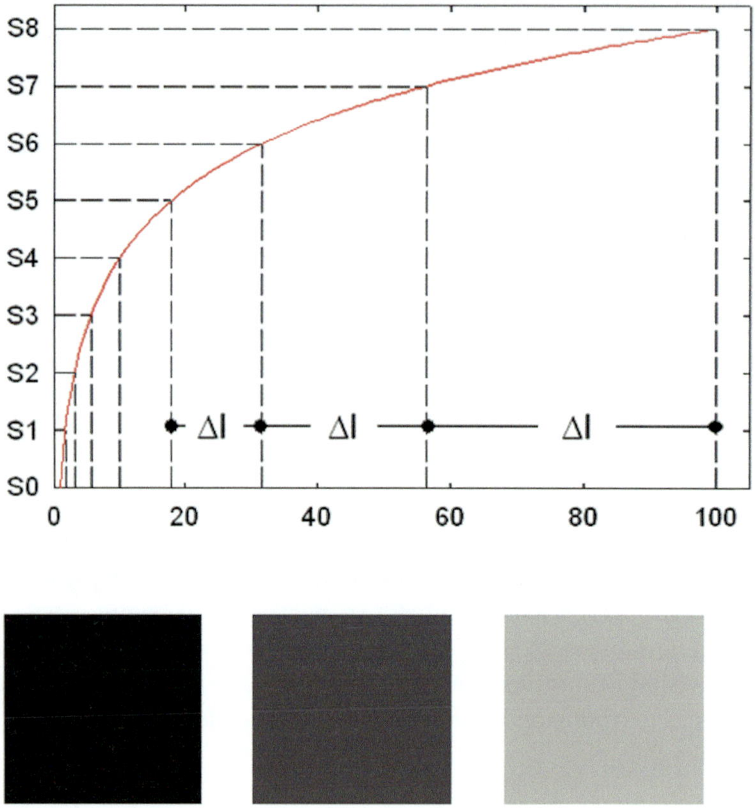

Figure 46. Graphic representations of the Fechner Law

Alternatively, Machine-Age scientists applied their perfect picture of 'ought' to design a better way of seeing. Three figures (Figures 47-49) presented by Judd represent the move from the quantification of colour vision to the standardization of the observer. Unlike most representations of the standard observer, these diagrams represent the curve along with the individual data points from which they were derived. Figure 48, in particular, compares the data from three studies, revealing just how artificial the standard observer is. The smoothness of the curve shows how the CIE scientists averaged the data in order to construct an ideal representation.

Luckiesh used the *optimum* point of the quantified visibility spectrum as the target towards which illuminating engineers should strive. He believed that lighting should use daylight as its model, not lamplight. Motivated to make electric light seem as natural as possible, he chose daylight as the baseline against which to measure artificial designs. He argued, 'As long

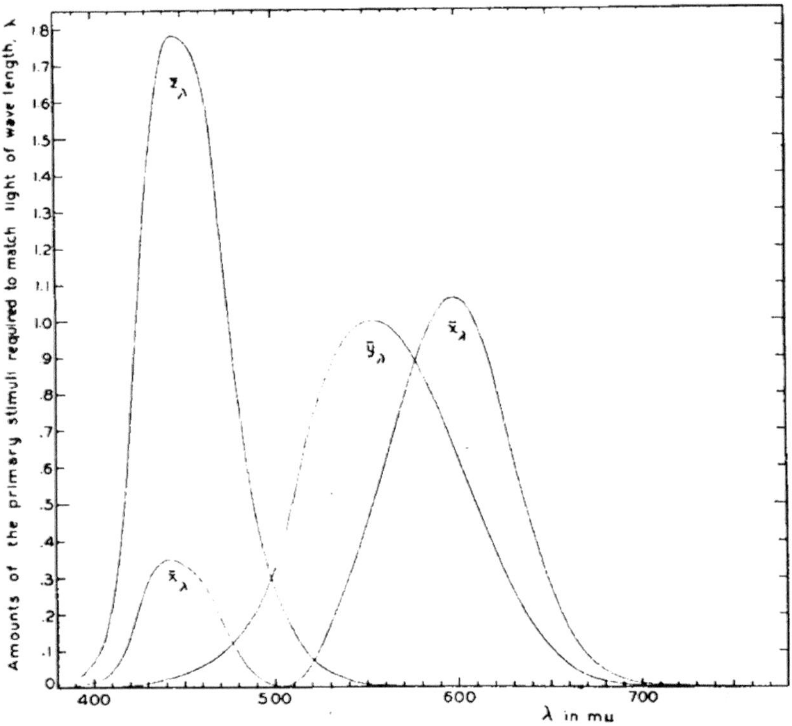

Figure 47. 'Color-matching functions for the 1931 CIE standard observer expressed in the CIE colorimetric coordinate system,' in Deane Judd, 'Fundamental Studies of Color Vision from 1860-1960,' *NAS Symposium* 55, 1966, 1326.

as study and consideration were confined to vision, practices were more or less anchored to the realm of barely seeing.[43] Illuminating engineers *ought to* be able to design better living conditions. Comparing the 'mere light' of nineteenth-century interiors to the more abundant lighting possibilities of electricity, he pushed for illuminating engineers to flood spaces with brightness.[44]

American periodicals during the interwar period popularized the notion of electric eyes as a metaphor for the power of science, the progress of industry, and the mastery of man over the natural world. While the rhetoric of space annihilation persisted, the tone had shifted. Instead of a sense of mastery over nature exemplified by the genius of a singular inventor (i.e.

43 Luckiesh, *Science of Lighting*, 4; Luckiesh, 'Human Seeing-machines,' 636; Cubitt, 139.

44 Luckiesh, 'New Science of Seeing,' 1.

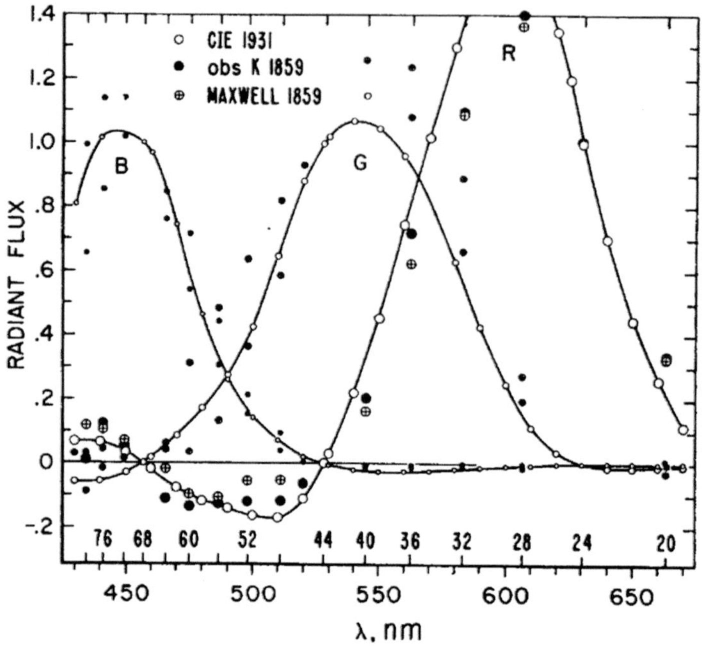

Figure 48. 'Radiant flux of the spectrum primaries,' in Deane Judd, 'Fundamental Studies of Color Vision from 1860-1960,' *NAS Symposium* 55, 1966, 1314.

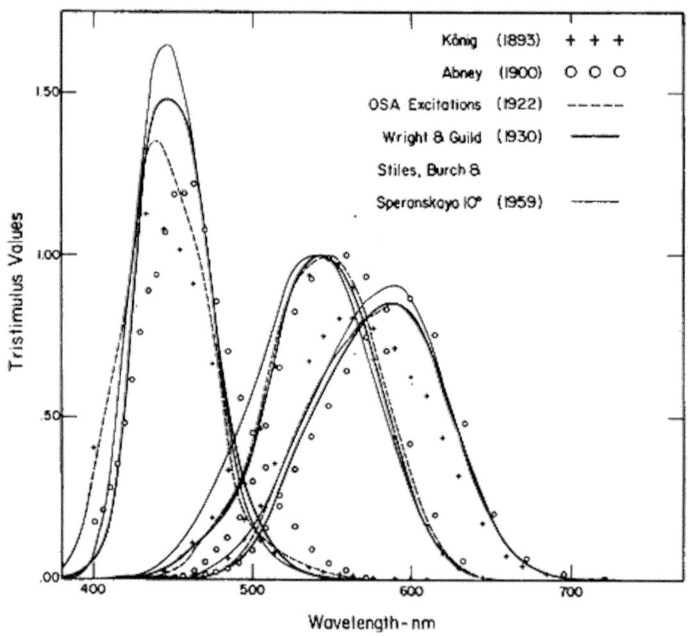

Figure 49. 'Color-matching functions expressed relative to the Young primaries,' in Deane Judd, 'Fundamental Studies of Color Vision from 1860-1960,' *NAS Symposium* 55, 1966, 1316.

The Wizard of Menlo Park), the interwar period embraced the ability of a society to operate like a well-oiled machine. The average citizen functioned as part of the mechanism. Popular-science periodicals promoted electric eyes as the metaphor for a better 'way of seeing', more efficient and more precise than the human eye.

Like the earlier hyperbole of annihilating space, the electric eye became a character unto itself, symbolizing the power of electronics, 'the mechanics of the infinitesimal and the improbable' to assist the human labourer.[45] Figure 50 shows a 1930 Westinghouse Electric advertisement from *Popular Science*, depicting 'Electric Eyes that never sleep'.[46] In popular-science periodicals and mainstream newspapers alike, electric eyes were shown to facilitate a modern way of life. They could be applied to tasks in the factory like colour grading and automation, and to everyday uses such as traffic monitoring, smoke detection, and remotely controlling doors.[47] Promoting electric eyes for all sorts of applications in industry and everyday life was a reminder that remote control, in all its connotations, was becoming a real possibility.

The discussion of the electric eye in the popular literature was overrun with trendy phrases, all of which emphasized the subject as a human seeing-machine: robot eyes, the artificial retina, machine vision given to 'Televox, the mechanical man'.[48] As one educational pamphlet explained, 'the photo-electric cell is a robot device with no brain behind it, and that its failures in the field of colorimetry are not the shortcomings of the photo-electric cell but of the operator who is making unreasonable demands upon it.'[49] The colourfully illustrated pamphlet made a point of noting the similarities between human and machine vision, and particularly the benefits that the machine offered

45 Caldwell, 'The Tiny Electron Works Wonders for Man.'

46 'Electric Eye that Never Sleeps,' (Westinghouse Ad) *Popular Science*, February 1930, 7.

47 While attempting to catalog every possible application would be fruitless, some of the most commonly suggested purposes include: smoke detection, colour grading, television camera transmission components, optical sound (sound-on-film), aiding astronomers in measuring the brightness of stars, automatic detection and controlling simple on/off functions such as light and motion detectors, automatic door openers, burglar alarms, traffic surveillance, factory conveyor belts and sorting functions, racetrack finish lines, and drinking fountains.

48 Robert Martin, 'Mystery Cell Aids Television,' *Modern Mechanics*, August 1930, 15-17; Arthur A. Stuart, 'Strange eyes that never sleep,' *Popular Science*, May 1929, 51; Robert Martin, 'Electric Eye Guides Ghostly Hands at a thousand jobs,' *Popular Science*, August 1933, 20-21; A. A. Hopkins, *'A Machine that is More than Human,' Scientific American 133, no. 6 (1925): 386-387*; Orestes Caldwell, 'Tireless Electric Eye Far Outdoes the Human,' *New York Times*, 16 July 1933; A. P. Peck, 'A Real "Electric Eye"', *Scientific American* 149, no. 3 (1933), 117.

49 Arthur Fawcett, *Electric Eyes: A Concise and Elementary Description of the Photo-electric Cell, for the Non-technical Reader: Its Uses in Industry, and Its Uses and Short-comings* (London: The Tintometer, 1954), 17.

Figure 50. 'Electric eye that never sleeps' (Westinghouse Ad), *Popular Science*, February 1930, 7

to overcome the flaws in human vision. A detailed graph on the inside cover introduces the parts of the spectrum visible to different types of animals. While humans can see a great swath along the middle, photoelectric cells are far more versatile. Periodicals emphasized the electric eyes as colour-blind, automatic, and mechanical; human vision was characterized as subjective, imprecise, and mercurial. As Olpin of Bell Labs explained it:

> In popular literature the photoelectric cell is frequently referred to as the 'electric eye' because it is commonly employed to do the work previously done by human observers. The response of the electrical eye to light of various colors however generally has been quite unlike that of the human eye. Of the photoelectric cells using pure metals as the light-sensitive element, only those employing caesium exhibit a response to colors that even roughly approximates that of the human eye.[50]

An accompanying diagram (Figure 51), similar to the pamphlet's illustration of the spectrum, depicts the sensitivity of different types of photoelectric

50 A. R. Olpin, 'New Types of Photoelectric Cells,' *Television News*, July 1931, 212.

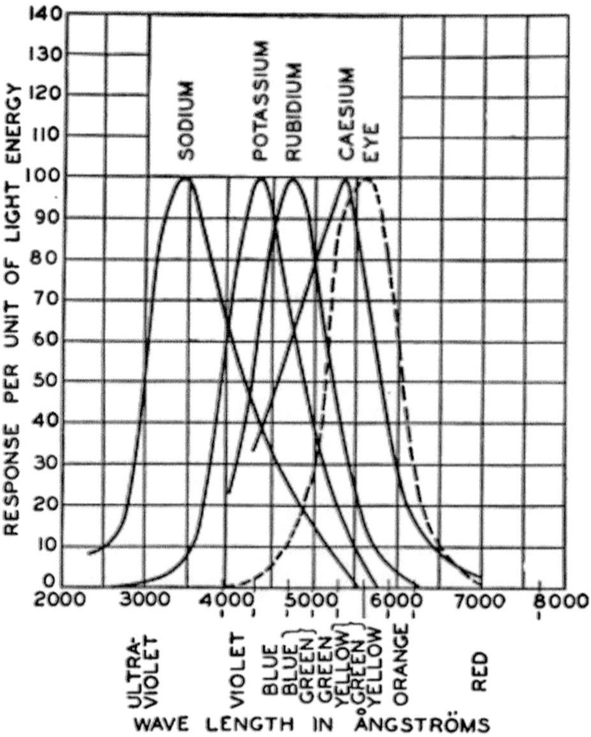

Figure 51. A. R. Olpin, 'New Types of Photoelectric Cells,' *Television News*, July 1931, 212-214.

Figure 52. E. E. Free, 'Magic Cell Works New Marvels: An Expert Tells How Photo-Electric "Eyes" Bring Television Nearer and Promise to Harness the Sun,' *Popular Science*, October 1928, 25-26.

cells. The literature makes clear that, while the caesium cells 'see' most similarly to human eyes, the rarity of the mineral makes it a less than ideal option. *Popular Science* and other magazines took pleasure in demonstrating the diversity of photoelectric cells (Figure 52).

Popular-science periodicals touted electric eyes as the hallmark of the Machine Age. If electric eyes were the viewer's mechanical counterpart, then, in a sense, they functioned as the mediation between vision and light. A decade before anyone had conceived of 'video' as an electronic medium, the language of television conveyed a visual process: the transmission and perception of light. In fact, because the scientific and popular languages of television diverged so radically, 'electric eyes' more commonly referred to television than any other term. Applying the metaphor of television as an extension of the eye, writers of popular science promoted electric eyes as a way of seeing, anthropomorphizing the photoelectric cell. Dunlap wrote in his book *Outlook for Television*: 'They have developed a new caesium photocell that "sees" red. It detects the red pigment of the skin and makes the image more lifelike.'[51] 'Electric eyes' provided a pervasive analogy that became synonymous with 'television'.

Herbert Ives, who devoted a great deal of effort to developing photoelectric cells while at Bell Labs, believed that it mattered less how the electric eyes 'see' than how they facilitated a better way of seeing for the human viewer.[52] Herbert Ives' design for the Bell Labs two-way television employed banks of photoelectric cells where one might expect to see a camera. The design arranged the cells in a mosaic behind specially tinted glass so as to capture a natural colour tone and gradient of light reflecting off the televised subject. Even though Ives directed the television project towards the goal of transmitting and receiving light values in the most efficient way possible, his practical approach demonstrates that he understood the human subject

51 Orrin Dunlap, *Outlook for Television* (New York: Harper and Brothers, 1932), 160. See also 'Radio-Images are Clarified,' *New York Times,* 11 January 1931.

52 Several obituaries note the importance of Ives' work in developing photocells. Buckley and Darrow; 'In Memoriam,' *The Reporter* (Bell Telephone Laboratories Magazine), December 1953: 23. Container 3, Frederic Eugene Ives and Herbert Eugene Ives Papers, Manuscript Division, Library of Congress, Washington, D. C.; Necrology, Physics Today, January 1954. Container 3, Frederic Eugene Ives and Herbert Eugene Ives Papers, Manuscript Division, Library of Congress, Washington, D.C.; 'Herbert E. Ives,' *Biographical Directory of American Men in Science, Part I: Physical Science*, Ninth Edition. Container 3, Frederic Eugene Ives and Herbert Eugene Ives Papers, Manuscript Division, Library of Congress, Washington, D. C.; 'H. E. Ives, Physicist, Succumbs,' *Montclair Times*, 19 November 1953; 'Dr. Herbert Ives, Pioneer in TV, Dies,' *New York Times*, 15 November 1953.

and the electric eyes as partners. The television viewer and the electric eye would work together, making it possible to 'see by electricity'.

Conclusion

The processes of perceptual accommodation resemble those of technological change. In fact, illuminating engineering suggests a way of thinking about perception and mediation as two sides of the same coin. Its interdisciplinary approach structures a dialogue between two seemingly incompatible ways of seeing the world: through subjective eyes (vision) and through objective optical science (light). That divide persists in academia today. Studies related to themes of perception and mediation can be found in many academic fields, two themes among many that can be investigated from interdisciplinary perspectives. A psychologist would research perception in a certain way, while an art historian would take a different approach. Psychologists, sociologists, media scholars, and historians of science and technology often deal with the same problems when investigating perception and technological mediation. Approaching the study of perceptual accommodation and technological mediation as a factor in cultural and historical change could provide the mutual ground on which interdisciplinary domains could learn to speak a common language. Sound studies and media archaeology are two such approaches, each embracing the interdisciplinary communication of knowledge and research in the investigation of questions of perception and technological mediation.

While the Machine-Age efficiency movement offers a fascinating window into a moment of cultural change that shaped 20th-century America, the story is still incomplete. This chapter has investigated the language and science of illuminating engineering that informed Machine-Age television. The next chapter fills in the gaps in the perception, reception, and experience of that technology. Illuminating engineering informed the design of Machine-Age television. Nowhere is that resemblance clearer than in Herbert Ives' two-way television project produced at Bell Labs. The final chapter takes readers into the television-telephone booth and questions what we think we know about what it means to watch or see through television.

Bibliography

Abramson, Albert. *The History of Television, 1880-1941.* Jefferson, NC: McFarland & Company, 1987.

Alexander, Jennifer. 'Efficiencies of Balance Technical Efficiency, Popular Efficiency, and Arbitrary Standards in the Late Progressive Era USA.' *Social Studies of Science* 38, no. 3 (2008): 323-349.

Alexander, Jennifer. *The Mantra of Efficiency: From Waterwheel to Social Control.* Baltimore, MD: Johns Hopkins University Press, 2008.

Beniger, James. *The Control Revolution: Technological and Economic Origins of the Information Society.* Cambridge, MA: Harvard University Press, 2009.

Boyce, Peter. *Human Factors in Lighting.* Boca Raton, FL. CRC Press, 2014.

Brill, Michael. 'How the CIE 1931 Color-Matching Functions were Derived from Wright-Guild Data.' *Color Research & Application* 23, no. 4 (1998): 259-259.

Buckley, Oliver and Karl K. Darrow. 'Herbert Eugene Ives (1882-1953).' Washington, DC: National Academy of Sciences, 1956.

Burns, R.W. 'Prophecy into Practice: The Early Rise of Videotelephony.' *Engineering Science & Education Journal* 4, no. 6 (1995): 33-40.

Cady, Francis and Henry Baldwin Dates, eds. *Illuminating Engineering.* New York: Wiley, 1925.

Caldwell, Orestes. 'The Tiny Electron Works Wonders for Man.' *New York Times,* July 6, 1930.

Caldwell, Orestes. 'Tireless Electric Eye Far Outdoes the Human.' *New York Times.* July 16, 1933.

Cochrane, Rexmond. *Measures for Progress: A History of the National Bureau of Standards.* Washington, D.C.: US Department of Commerce, 1966.

Codman, John. 'Advent of Illuminating Engineering.' In *Technology and Industrial Efficiency.* New York: McGraw-Hill, 1911.

Corwin, Sharon. 'Picturing efficiency: Precisionism, Scientific Management, and the Effacement of Labor.' *Representations* 84, no. 1 (2003): 139-165.

Cubitt, Sean. *The Practice of Light: A Genealogy of Visual Technologies from Prints to Pixels.* Cambridge, MA: MIT Press, 2014.

Danziger, Kurt. *Constructing the Subject: Historical Origins of Psychological Research.* Cambridge, UK: Cambridge University Press, 1994.

Davis, Lennard J. 'Constructing Normalcy: The Bell Curve, the Novel, and the Invention of the Disabled Body in the Nineteenth Century.' In *The Disability Studies Reader,* 2nd edition, edited by Lennard J. Davis, 3-16. New York: Taylor & Francis, 2006.

Dunlap, Orrin. *Outlook for Television.* New York: Harper and Brothers, 1932.

Electrical Review and Western Electrician. 'Physical Laboratory of the National Electric Lamp Association: A Laboratory for Research in Illumination and Electric Lighting' (reprinted in *Science,* December 24, 1909). September 10, 1910, 507-511.

Ernst, Waltraud. 'The Normal and the Abnormal: Reflections on norms and normativity.' In *Histories of the Normal and the Abnormal: Social and Cultural*

Histories of Norms and Normativity, edited by Waltraud Ernst, 1-25. New York: Routledge, 2006.

Fawcett, Arthur. *Electric Eyes: A Concise and Elementary Description of the Photo-electric Cell, for the Non-technical Reader: Its Uses in Industry, and Its Uses and Short-comings*. Place: The Tintometer, 1954

Good Lighting and Illuminating Engineer. 'A Year's Progress in Illuminating Engineering.' January 1912, 2.

Haber, Samuel. *Efficiency and Uplift: Scientific Management in the Progressive Era, 1890-1920*. Chicago, IL: University of Chicago Press, 1964.

Hays, Samuel P. *Conservation and the Gospel of Efficiency: The Progressive Conservation Movement, 1890–1920*. Cambridge, MA: Harvard University Press, 1959.

'Herbert E. Ives,' *Biographical Directory of American Men in Science*, Part I: Physical Science, Ninth Edition. Container 3, Frederic Eugene Ives and Herbert Eugene Ives Papers, Manuscript Division, Library of Congress, Washington, D.C.

Hirshberg, Leonard Keene. 'Why Our Eyes Very Often Deceive Us.' Newspaper clipping, *Evening Sun*, July 20, 1914, Container 4, Frederic Eugene Ives and Herbert Eugene Ives Papers, Manuscript Division, Library of Congress, Washington, D.C.

Hopkins, A.A. 'A Machine that is More than Human.' *Scientific American*. December 1925, 336-7.

Hunter, Richard and Richard W. Harold. *The Measurement of Appearance*. New York: Wiley, 1987.

IES Lighting Handbook: A Standard Lighting Guide. New York: Illuminating Engineering Society, 1947.

The Illuminating Engineer, 'The Johns Hopkins University Course of Lectures on Illuminating Engineering,' July 1910, 256-7.

Induction Letter, Illuminating Engineering Society, September 25, 1908. Container 9, Frederic Eugene Ives and Herbert Eugene Ives Papers, Manuscript Division, Library of Congress, Washington, D.C.

Ives, Herbert. 'An Illuminating Engineer's Conception of an Ideal Light.' *Transactions of the American Electrochemical Society* 17 (1915): 419-433.

Ives, Herbert and E.F. Kingsbury. 'The Applicability of Photoelectric Cells to Colorimetry.' *Journal of the Optical Society of America* 20 (September 1930): 541-563.

Johnston, Sean. 'The Construction of Colorimetry by Committee.' *Science in Context* 9 (1996): 387-420.

Johnston, Sean. *A History of Light and Colour Measurement: science in the shadows*. Bristol, UK: Institute of Physics Publishing, 2001.

Jolley, Leonard, John Waldram and George Wilson. *The Theory and Design of Illuminating Engineering Equipment*. New York: Wiley, 1931.

Jones, Lloyd. 'The Historical Background and Evolution of the Colorimetry Report of the OSA.' *Optometry and Vision Science* 21 (August 1944): 317-333.

Jordan, John M. *Machine-Age Ideology: Social Engineering and American Liberalism, 1911-1939*. Chapel Hill: University of North Carolina Press, 1994.

Judd, Deane. 'The 1931 I. C. E.: Standard Observer and Coordinate System for Colorimetry.' *Journal of the Optical Society of America* 23 no. 10 (1933): 359-373.

Judd, Deane. 'Fundamental Studies of Color Vision from 1860-1960.' *NAS Symposium* 55 (1966): 1313-1330.

Judd, Deane and Günter Wyszecki. *Color in Business, Science and Industry*. New York: Wiley, 1963.

Lane, Alfred. 'The Real Facts About Television.' *Popular Science Monthly,* September 1928, 43-44.

Light: Its Use and Misuse. New York: Illuminating Engineering Society, 1912.

Luckiesh, Matthew and Frank Moss. 'The Human Seeing-Machine.' *Journal of the Franklin Institute* 215 (June 1933): 629-654.

Luckiesh, Matthew and Frank Moss. 'The Motion Picture Screen as a Lighting Problem.' *Journal of the Society of Motion Picture Engineers* 26 (May 1936): 578-591.

Luckiesh, Matthew and Frank Moss. *The New Science of Lighting.* Schenectady, NY: Lighting Research Lab, General Electric, 1934.

Luckiesh, Matthew and Frank Moss. 'The New Science of Seeing.' *Transactions of the Illuminating Engineering Society* 25 (January 1930): 15-39.

Luckiesh, Matthew and Frank Moss. *The Science of Seeing.* Place: D. Van Nostrand Co, 1937.

Martin, Robert. 'Electric Eye Guides Ghostly Hands at a Thousand Jobs.' *Popular Science,* August 1933, 20-21.

Martin, Robert. 'Mystery Cell Aids Television.' *Modern Mechanics,* August 1930, 15-17.

Mills, Mara. 'Deafening: Noise and the Engineering of Communication in the Telephone System.' *Grey Room* 43 (2011): 118-143.

Mindell, David A. *Between Human and Machine: Feedback, Control, and Computing Before Cybernetics.* Baltimore, MD: Johns Hopkins University Press, 2002.

Montclair Times. 'H. E. Ives, Physicist, Succumbs.' November 19, 1953.

Mumford, Lewis. *Technics and Civilization.* New York: Harcourt, 1934.

Nason, C.H.W. 'Photocell Applications in Projection.' *Projection Engineering,* April 1930, 15-17.

New York Times. 'Dr. Herbert Ives, Pioneer in TV, Dies.' November 15, 1953.

New York Times. 'Radio-Images are Clarified.' January 11, 1931.

Nye, David. *Electrifying America: Social Meanings of a New Technology, 1880-1940.* Cambridge, MA: MIT Press, 1992.

Olpin, A. R. 'New Types of Photoelectric Cells.' *Television News,* July 1931, 212.

Otter, Chris. *The Victorian Eye: A Political History of Light and Vision in Britain, 1800-1910.* Chicago: University of Chicago Press, 2008.

Peck, A.P. 'A Real `Electric Eye'.' *Scientific American*, September 1933, 117.

Physics Today, Necrology, January 1954. Container 3, Frederic Eugene Ives and Herbert Eugene Ives Papers, Manuscript Division, Library of Congress, Washington, D.C.

Popular Science. 'Electric Eye that Never Sleeps' (Westinghouse Ad). February 1930, 7.

Projection Engineering. 'The Photoelectric Cell: Light Energy and the Photoelectric Effect.' November 1929, 35.

Reeb, O. 'A Consideration of the Screen Brightness Problem.' *Journal of the Society of Motion Picture Engineers* 32 (May 1939): 485-494.

The Reporter (Bell Telephone Laboratories Magazine). 'In Memoriam,' December 1953, 23. Container 3, Frederic Eugene Ives and Herbert Eugene Ives Papers, Manuscript Division, Library of Congress, Washington, D.C.

Rubinstein, Charles. 'Optics at Bell Laboratories – General Optics, television, and Vision.' *Applied Optics* 11, no. 11 (1972): 2401-2411.

Rydell, Robert. *World of fairs: the century-of-progress expositions.* Chicago: University of Chicago Press, 1993.

Saunders, Barbara and Jaap Van Brakel. 'The Trajectory of Color.' *Perspectives on Science* 10, no. 3 (2002): 302-355.

Schanda, Janos, ed. *Colorimetry: Understanding the CIE System.* New York: Wiley, 2007.

Sorin-George, Toma, Ana-Maria Grigore, and Paul Marinescu. 'The Emergence of Scientific Management in America.' *Manager* 19 (2014): 128-131.

Sterne, Jonathan. *MP3: The Meaning of a Format.* Durham, NC: Duke University Press, 2012.

Sterne, Jonathan and Dylan Mulvin. 'The Low Acuity for Blue: Perceptual Technics and American Color Television.' *Journal of Visual Culture* 13, no. 2 (2014): 118-138.

Stuart, Arthur A. 'Strange eyes that never sleep.' *Popular Science,* May 1929, 51.

Thompson, Kenneth. 'Introduction to the Early Sociology of Management and Organizations.' In *Scientific Management.* New York: Routledge, 2003.

Winkler, Stefan. *Digital Video Quality: Vision Models and Metrics.* New York: John Wiley & Sons, 2005.

Wu, Hong Ren and Kamisetty Ramamohan Rao, eds. *Digital Video Image Quality and Perceptual Coding.* Boca Raton: CRC press, 2005.

Wurtzler, Steve. *Electric Sounds: Technological Change and the Rise of Corporate Mass Media.* New York: Columbia University Press, 2007.

5. The Ikonophone

Bell Laboratory's Two-Way Television Project

Abstract

Chapter Five presents a case study of Bell Labs's two-way television project, also called the 'Ikonophone'. Looking closely at press coverage of the Ikonophone project and the language journalists, writers of popular science, and engineers used to describe their experiences interacting with the screen, this chapter examines the reception and construction of television in the 1920s. Engineers described the method of designing an environment in which the user was made to feel as if they were face-to-face with the distant party. Witnesses responded to what they saw on the screen with a confused combination of metaphors, mingling expressions of a feeling of closeness with reactions to its uncanny artificiality. The Ikonophone marks the moment of television's transition from a technology into a visual medium.

Keywords: Screen-mediated communication; Bell Labs; Machine Age; mechanical television

When the first English-language monograph devoted entirely to the subject of television appeared on bookshelves in 1926, it indicated the burgeoning of a craze.[1] Along with the help of writers of popular science who published news, educational, promotional, and opinion pieces on the subject, the

1 Alfred Dinsdale, *Television: Seeing by Wireless* (London: Sir I. Pitman & Sons, 1926). See also Thomas Baker, *Wireless Pictures and Television* (New York: Nostrand, 1927); Edgar T. Larner, *Practical Television* (New York: Nostrand, 1928); Sydney Moseley, *Television Today and Tomorrow* (London: Isaac Pitman, 1929); H. H. Sheldon and Edgar Norman Grisewood, *Television: Present Methods of Picture Transmission* (New York: Nostrand, 1929); Raymond F. Yates, *ABC of Television; Or, Seeing by Radio* (New York: Chapman & Hall, 1929); C. F. Jenkins, *Radiomovies, Radiovision, Television* (Washington, D.C.: National Capital Press, Inc., 1929); Edgar H. Felix, *Television, Its Methods and Uses* (New York: McGraw-Hill Book Company, Inc., 1931); Orrin Dunlap, *Outlook for Television* (New York: Harper & Brothers Publishers, 1932).

Roberts, I., *Visions of Electric Media: Television in the Victorian and Machine Ages.*
Taylor & Francis Group, 2019
DOI 10.5117/9789462986596_CH05

evolving discourse helped to construct expectations for television as both a type of visual perception and a type of mass medium. But those expectations constructed in the press differed in important ways from the firsthand accounts offered by television's first viewers. Visitors to the Bell Laboratory demonstration of the Ikonophone (1927-1930) vacillated between describing *looking through* a window and *looking at* a screen. This chapter examines the sharp contrast between the popular science of television and the experiences of its earliest users. It focusses on the popularized conception of television that developed in the late 1920s, and how it differed from the nineteenth-century rhetoric of space annihilation in which 'seeing by electricity' had been steeped.

These periodicals introduce a new character along with the new machine, called the 'looker-in', the visual counterpart to 'listening in' to a radio broadcast.[2] Characterizing the television viewer in this way supported an understanding of television as a window that provided unmediated access to another world or distant place. But, when witnessing firsthand, viewers struggled to negotiate between the impression of the screen as a visual *representation* and the 'uncanny' recognition that, in the case of two-way television, the person on the screen could also look back.[3] The public demonstrations of the Bell Labs two-way television project in the late 1920s illustrate this conflict, as the character, purpose, and potential of the medium coalesced across the pages of the American popular press.[4]

Bell Labs's two-way television employed a mechanical-optical approach that contrasted with the all-electronic method exemplified by cathode ray tube screens. The Ikonophone implemented a combination of a photoelectric

2 For example, *Television News* announced the winner of the 'New Word Contest' to replace the inadequate 'Looker-In' to designate the role of the television viewer. 'Results of $50.00 "New Word" Contest,' *Television News,* July-August 1931, 211; 'Remarkable Remarks,' *The Independent,* 11 October 1928, 2. See also Orrin Dunlap, *Outlook for Television,* 141, 174; James Miller, 'The Latest in Television,' *Popular Mechanics,* September 1929, 474; Alfred Goldsmith, 'Electrical Entertainment: A Glimpse of the Future,' *New York Times,* 22 March 1931; 'Latest Television Broadcast Station,' *Everyday Science and Mechanics,* November 1931, 690.

3 Dunlap, *Outlook,* 107; R. L. Duffus, 'The Magic of a Vast Laboratory,' *New York Times,* 8 June 1930; Laurence Cockaday, 'The Latest Developments in Television Methods,' *Radio Craft,* July 1930, 22-24; '100 Trade Leaders Tour Laboratories,' *New York Times,* 8 October 1930.

4 See, for example, Steven Shapin and Simon Shaffer, *Leviathan and the Air-Pump: Hobbes, Boyle and the Experimental Life* (Princeton, NJ: Princeton University Press, 1985). In Chapter 2, 'Seeing and Believing,' the authors discuss the role of technologies (material, literary, and social) that work to mediate and legitimate knowledge between scientist and community. Shapin and Shaffer call this 'virtual witnessing'.

cell, a Nipkow disc, and a neon lamp.[5] On the transmitting end, capture light reflects off an object with a photoelectric cell. The photoelectric cell transduces (or converts) that light into an electric current. On the receiving end, another photoelectric cell transduces the electric current back into light. A lamp turns the current into visible light. A Nipkow disc spinning in front of the lamp synchronizes with the transmitting station to reconstruct a visible image perceivable to the eye. A lens can be employed to focus the light.

Historians have described mechanical television as 'a curiosity'.[6] The history of mechanical television has received much more attention by European media scholars, and much of this work has yet to be translated into English, much less language accessible to non-technical readers.[7] In the words of Dutch media historians Van Ende, Ravesteijn, and De Wit:

> There were once important incentives supporting mechanical television in its rivalry to electronic television. The electronic television gained preeminence not because of purely technical considerations, as is so often assumed, but also because of social ones.[8]

Several obstacles stand in the way of conveying the media history of mechanical television. The prominence of technological determinist approaches leads to the tendency to assume that mechanical television was a dead end. Resurrecting the sense of contingency and possibility prevalent at the time

5 For technical details on the construction of mechanical television systems, see R. W. Burns, 'Prophecy into Practice: The Early Rise of Videotelephony,' *Engineering Science & Education Journal* 4, no. 6 (1995): 33-40; Donald McLean, *Restoring Baird's Image* (New York: Wiley-IEE, 2000).

6 Jan Van den Ende, Wim Ravesteijn, and Dirk De Wit. 'Shaping the Early Development of Television,' *Technology and Society Magazine, IEEE* 16, no. 4 (1997): 13.

7 R. W. Burns, 'The Contributions of the Bell Telephone Laboratories to the Early Development of Television,' *History of Technology* 13 (1991): 181-213; George Shiers, 'SMPTE Historical Note: The Rise of Mechanical Television, 1901–1930,' *SMPTE Journal* 90, no. 6 (1981): 508-521; Charles Rubinstein, 'Optics at Bell Laboratories – General Optics, television, and Vision,' *Applied Optics* 11, no. 11 (1972): 2401-2411; Donald McLean, *Restoring Baird's image* (London: IEE, 2000); Antonio Perez Yuste, 'La televisión mecánica (The Mechanical Television),' in *Detrás de la Cámara. Historia de la Televisión y de sus Cincuenta Años en España* (*Behind the Scenes: History of Television and its Fifty Years in Spain*) (Madrid: Colegio Oficial de Ingenieros de Telecomunicación, 2008), 65-82; Franz Pichler, 'Mechanisches Fernsehen: Fernsehempfang in der vorelektronischen Zeit (Mechanical Television: Reception in the pre-electornic days,' *Plus Lucis* 2 (2001): 21-26; Tina Zeise, *Geschichte und Technik des analogen Fernsehens* (*History and technology of Analog Television*) (Munich: GRIN Verlag, 2006).

8 Jan Van den Ende, Wim Ravesteijn, and Dirk De Wit, 'Shaping the early Development of Television,' *Technology and Society Magazine, IEEE* 16, no. 4 (1997): 13.

of these demonstrations presents a challenge to the historical imagination. Seeing as how the mechanical-optical approach represents a method long since abandoned, those hurdles in explaining how it worked and what it looked like come back full force, complicating the media history of early television.

This chapter examines the results of the Bell Labs television demonstrations as they speak to the construction of medium identity in design, technique, and practice. This chapter begins with a profile of the Bell Labs two-way television project. The apparatus' unusual design and engineering methodology offer a look into a very different way of going about solving the problem of television in contrast to the CRT and flat-screen displays that became the standards. The second part examines representations of television in the late 1920s popular-science press. The television craze inspired contributors to work towards a definition of television as a medium, an industry, and a practical reality. Technical descriptions of electrical currents contrasted with attempts to explain television as a special kind of seeing. Examining the many ways writers found to express the concept of television reveals the intricate way the medium's identity began to take shape. The third part turns to those who witnessed the television demonstrations, paying particular attention to the way their descriptions of the experience contrast with those of the technicians and engineers. The efforts of engineers to design a television-telephone booth that would simulate the feeling of intimacy experienced in a face-to-face encounter was met with mixed reactions. The common assumption that the television offered *a window through* which the users communicated mingled with the impression of a *representation depicted* on the screen.[9] Read together, these two types of reactions exemplify the conflict between realism and illusionism that goes back to the early days of cinema. Finally, by comparing and contrasting the rhetoric that emerged with the television craze with the cultural constructions of 'seeing by electricity' in the nineteenth century, the chapter concludes by raising questions about the cultural and technological constructions of media.

Of all the models of television presented in the first decade of its operation, the Bell Labs two-way television resembles the Victorian prophecy of the Telephonoscope most closely. Also promoted under the name of 'Ikonophone', as preferred by its project director, Dr. Herbert Ives, it continues the legacy of 'television' that began with Edison, Senlecq, Szczepanik, and others.[10]

9 Anne Friedberg, *The Virtual Window: from Alberti to Microsoft* (Cambridge, MA: MIT Press, 2006).

10 *Television*, reprinted from *Bell Laboratories Record*, June 1927, 22.

The first such machine not only to take material form but also to display for the viewer an actual, reliable, and identifiable image, Ives' Ikonophone, was recognized, at least by the Bell community, as the birth of television.[11] If the Telephonoscope survived as the dream of television, the Ikonophone represents its reality. Examining the similarities and differences between the conceptions of television condensed in these two images reveals the deeply embedded assumptions about the identity of television and the modern audiovisual media.

It was hailed as many things:

- 'The menace of television [...] haunted by the lurking perils of snap-shots'[12]
- 'The conquest of nature'[13]
- 'It seems like magic to the watchers[...]'[14]

These assumptions may in fact have more to do with culture and history than they do with the technologies themselves. Assumptions about audiences and modes of communication, mediation, and the culturally constructed limits of the human body come through consistently in the late nineteenth- and early 20th-century popular science of television. While the messages remain the same, both ages infuse television discourse with a distinct tone. Contrasting the late nineteenth-century and early 20th-century versions of television condensed in the images of the telephonoscope and the Ikonophone will reveal the many embedded assumptions about medium identity and the cultural and technological construction of experience.

Very little attention has been paid to Machine-Age television in recent scholarship. In the limited scholarship on Bell Labs's early television experiments, scholars gravitate towards questions of success and failure on a commercial and industrial scale.[15] Studies on videotelephony, the history of Bell Labs, and early television make up the bulk of research.[16] Historians of

11 Herbert Ives, 'Television: 20th Anniversary,' *Bell Laboratories Record* 25, no. 5 (May 1947): 190-193; 'Birth of Television: TV Marks its Birthday,' *The Reporter*, April 1957, 4-6, 14.

12 Edward Van Zile, 'New Perils of Television,' *The Spur*, 1 May 1928, 41.

13 'Far-off Speakers Seen as Well as Heard Here in a Test of Television: Like a Photo Come To Life,' *New York Times*, 8 April 1927.

14 'Television Now Reality; Device Demonstrated,' *Troy Record*, 8 April 1927.

15 R. W. Burns, 'The Contributions of the Bell Telephone Laboratories to the Early Development of Television,' *History of Technology* 13 (1991): 181-213.

16 Abramson, *History of Television, 1880-1941*; Burns, 'Contributions'; R. W. Burns, 'Prophecy into practice: the early rise of videotelephony,' *Engineering Science & Education Journal* 4, no. 6 (1995), 33-40; Charles Rubinstein, 'Optics at Bell Laboratories – General Optics, television, and Vision,' *Applied Optics* 11, no. 11 (1972), 2401-2411; George Shiers, 'The Rise of Mechanical Television,

technology tend to treat projects of this sort as dead ends.[17] But there is much more to be learned from studying the Ikonophone. Combining cultural history and media theory, a method common for media archaeologists, locates the Ikonophone as a product of its context.[18] Political, scientific, and technological forces shaped Machine-Age culture. Social, human, and illuminating engineering inform the way Bell Labs designed the Ikonophone. The Machine Age offers a wealth of material for the media archaeologist, as the technology of this period presents so many roads not taken. It also marks a watershed of possible outcomes, intermedial combinations, and rich collaborations between media, entertainment, communications, and industry.[19]

Designing Television

On 7 April 1927, Bell Laboratories invited a small group of representatives from the press to witness a special demonstration of two-way television.[20] The goal of the presentation was to show how the experimental equipment could broadcast on both wired and wireless channels. They broadcast Herbert Hoover, then Secretary of Commerce, in picture and voice from Washington, D. C.. His words came over the loudspeaker to the New York crowd, while they watched him speak on a big screen. Hoover's address captured the monumental nature and sense of contingency in the moment:

1901–1930,' *SMPTE Journal* 90, no. 6 (1981), 508-521; Jon Gertner, *The Idea Factory: Bell Labs and the Great Age of American Innovation* (New York: Penguin, 2012).

17 Jonathan Coopersmith, *Faxed: The Rise and Fall of the Fax Machine* (Baltimore, MD: John Hopkins University Press, 2015); Kenneth Lipartito, 'Picturephone and the Information Age: The Social Meaning of Failure,' *Technology and Culture* 44, no. 1 (2003): 50-81; Jennifer Light, 'Facsimile: A Forgotten "New Medium" from the 20th Century,' *New Media & Society* 8, no. 3 (2006): 355-378.

18 Jussi Parikka, *What is Media Archaeology* (New York: John Wiley & Sons, 2013); Erkki Huhtamo and Jussi Parikka, 'Introduction: An Archaeology of Media Archaeology,' in *Media archaeology: Approaches, applications, and implications* (Berkeley, CA: University of California Press, 2011): 1-26; Anne-Katrin Weber, 'Recording on Film, Transmitting by Signals: The Intermediate Film System and Television's Hybridity in the Interwar Period,' *Grey Room* 56 (2014): 6-33; Grant Wythoff, 'Pocket Wireless and the Shape of Media to Come, 1899–1922,' *Grey Room* 51 (2013): 40-63; Mara Mills, 'The Audiovisual Telephone: A Brief History,' in *Handheld? Music Video Aesthetics for Portable Devices*, ed. Henry Keazor (Heidelberg: ART-Dok, 2012), 34-47.

19 Steve Wurtzler, *Electric Sounds*.

20 A. A. Albelli, 'What's Next in Television?' *Popular Mechanics*, July 1927; 'Far-off Speakers Seen as Well as Heard Here in a Test of Television: Like a Photo Come to Life,' *New York Times*, 8 April 1927; 'Washington Hails The Test: Operator There Puts Through the Calls as Scientists Watch,' *New York Times*, 8 April 1927.

Figure 53. Front of the 1927 model. Photo courtesy of the Early Television Museum, Hillard, OH.

'Human genius has now destroyed the impediment of distance in a new respect, and in a manner hitherto unknown. What its uses may finally be, no one can tell, any more than man could foresee in past years the modern development of the telegraph and the telephone.'[21] Next, a vaudeville act was broadcast over the airwaves from a radio station in Whippany, New Jersey, which included a minstrel show and a song-and-dance routine.

A large screen at right displayed the transmissions from Washington and Whippany. The apparatus facilitated a television-telephone conversation between Hoover and AT&T President Walter Gifford. A candlestick telephone sat on a shelf and the picture came through on a small display mounted in the wooden panel. The picture frame-like display presented the viewer with what might appear to be a screen or window. However, the wooden apparatus served merely as the casing for the mechanism (Figure 53). This

21 *Television*, reprinted from *Bell Laboratories Record*, June 1927, 6. See also 'Program of Demonstration,' (7 April 1927), Container 3, Frederic Eugene Ives and Herbert Eugene Ives Papers, Manuscript Division, Library of Congress, Washington, D. C.; 'Preliminary Demonstration Program' for 7 April 1927 Television demonstration, Container 8, Frederic Eugene Ives and Herbert Eugene Ives Papers, Manuscript Division, Library of Congress, Washington, D.C.

initial version of the Ikonophone had no 'screen' in the conventional sense. The display served as a part of the casing, and a lens inside the machine focussed the light. The viewer actually looked through a kind of peephole. As one writer put it, 'the light source is the picture itself'.[22]

Using technical language, scientific periodicals spoke of photoelectric cells and Nipkow discs. Explanations of mechanical-optical television transmission systems generally consisted of the process by which light was captured and transmitted across electrical circuits. The mechanical-optical approach to television transmission was explained as follows:

> The process employed in the 1927 demonstration of television over electrical circuits involved an intense beam of light, which scanned the object. Reflected light was picked up by a bank of large photo-electric cells, and converted into variations of electric current. Sufficiently amplified, this current controlled the brightness of a neon lamp at the receiving station. The neon lamp when scanned by a moving aperture in synchronization in with the initial beam of light appeared to the observer to re-create the original object.[23]

As represented in illustrations and depictions in the popular science (see Chapter Three), at its most basic, the mechanical-optical approach adopted by Ives involved a meticulously engineered system. A photoelectric cell at the transmitting end detects light and converts it into electrical current; on the other end, a lamp receives that pulse and turns the energy back into visible light; a spinning disc placed in front of the light synchronizes with a mechanism at the transmitting end to reconstruct the pulses into a perceivable image; a lens located in front of the disc focusses the light so that the viewer can better see 'the light flashing before their eyes'. As if that was not enough, the process was complicated in two-way television by the fact that both transmitting and receiving stations were modular; the viewer was also the subject. The booths included both 'camera' and 'screen'.

The technical nature of the Ikonophone presented the technician with several obstacles when presenting to a general audience. The mechanical-optical television system, as its advocates conceded, were 'hard for a lot of people to understand'.[24] One might have expected to find a screen and

22 James Miller, 'The Latest In Television,' *Popular Mechanics,* September 1929, 473.

23 *Two-Way Television and A Pictorial account of its background* (AT&T/Bell Laboratories, 1930), 5.

24 C. F. Jenkins, 'Life Size Radio Movies are coming,' *Modern Mechanics and Inventions,* May 1930, 72.

a camera, but the Ikonophone included neither of these features. When attempting to explain what the viewer saw in the visual apparatus, technicians privileged the physical view, relying mostly on language pulled from optics and physiology. Looking into a camera's lens provides a focal point, a place for the eyes to fall with the expectation of being photographed. It offers a sense of connection between the here and the now and what one would imagine to exist on the other end.

In mechanical-optical systems, the image only existed when there was a viewer watching, like the falling tree that makes no sound when no one is around to hear it. Herbert Ives and Charles F. Jenkins, who both pioneered the mechanical-optical approach, took the stance that the television image was a subjective perception. Ives described it, in the classic fashion of a physicist specialized in optics, using language as precise as he could muster. In an official communique to the Bell Systems community, he wrote: 'Viewed through the aperture of the metal plate, the holes in the disc form moving spots of varying brightness which paint a picture on the retina of the eye.'[25] Jenkins explained it in a way that may have been easier for a general audience to understand:

> 'There is no picture except what your eyes and brain form. You think you see a picture in the machine, but all you really see is a rapidly fluctuating point of light.' [...] But actually neither pictures [live television image or broadcast pre-recorded movie] exists outside your brain [....] The [light] fluctuations paint bright, shadowy and dark spaces, and when the eye assembles them as a whole you see a picture.[26]

These descriptions supported a sense that television was simply a process by which the light reflecting off an object was relayed to a distant viewer. It gave the impression that there was nothing technological mediating the experience, but that it was rather an electrical transmission of something that equated to vision.

Following the demonstration in April 1927, the television team went back to the drawing board to iron out the kinks. Three years later, in April 1930, they unveiled an upgraded model, and this time invited the public to

25 *Two-Way Television and A Pictorial account of its background* (New York: AT&T/Bell Laboratories, 1930), 7.

26 C. F. Jenkins, 'Life Size Radio Movies are coming,' *Modern Mechanics and Inventions,* May 1930, 72-73.

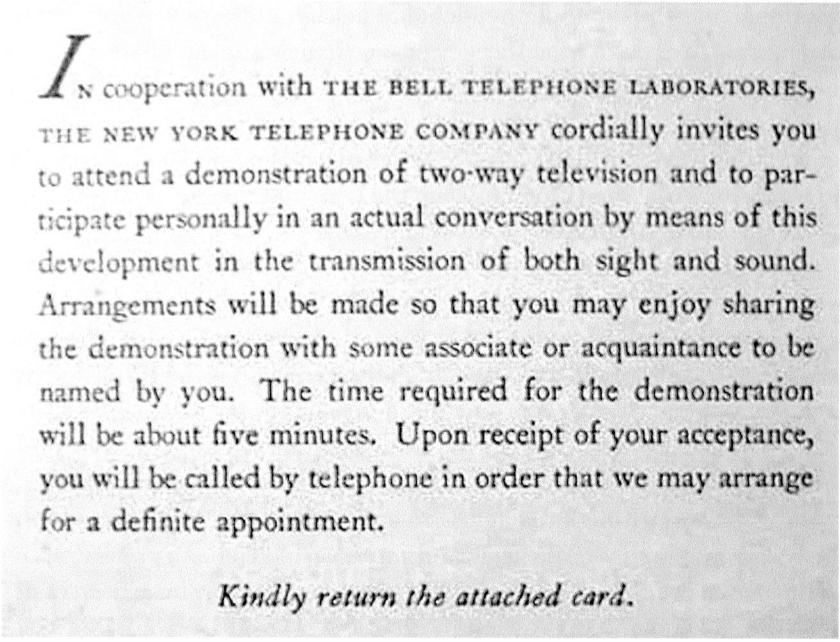

IN cooperation with THE BELL TELEPHONE LABORATORIES, THE NEW YORK TELEPHONE COMPANY cordially invites you to attend a demonstration of two-way television and to participate personally in an actual conversation by means of this development in the transmission of both sight and sound. Arrangements will be made so that you may enjoy sharing the demonstration with some associate or acquaintance to be named by you. The time required for the demonstration will be about five minutes. Upon receipt of your acceptance, you will be called by telephone in order that we may arrange for a definite appointment.

Kindly return the attached card.

Figure 54. Television Demonstration Invitation Card. Photograph courtesy of the Early Television Museum, Hillard, OH.

see it for themselves over a year-long experimental trial.[27] It was a series of invitation-only events targeted at garnering cultural capital from the press, industry representatives, businessmen, and social elites. Figure 54 presents one such invitation. One side included the instructions, and the reverse listed the date, time, and location of the event.

27 'Two-Way Television Closes Two-Mile Gap,' *Washington Post,* 9 April 1930; '2-Way Television in Phoning Tested,' *New York Times,* 10 April 1930; T. R. Kennedy, 'Speakers on Phone See Images of Each Other,' *New York Times,* 13 April 1930; 'Ethereal Images Baffle Experts,' *New York Times,* 20 April 1930; 'Two-Way Television,' *The Scientific Monthly* 30, no. 5 (May 1930): 476-480; Orrin Dunlap, 'Television Brought Nearer the Home,' *New York Times,* 25 May 1930; R. L. Duffus, 'The Magic of a Vast Laboratory,' *New York Times,* 8 June 1930; Art Brown, 'Television is Ready for Business,' *Nation's Business,* June 1930, 47; 'Deaf "Hear" 2 Miles in Television Test,' *New York Times,* 3 July 1930; 'Talk, Hear, SEE on This Phone: Two-Way Television Is Demonstrated in Laboratory As an Engineering Stunt,' *Popular Science,* July 1930, 22, 123; Robert Martin, 'Mystery Cell Aids Television,' *Modern Mechanics,* August 1930, 15-17, 119; 'Television Now Gives Radio eyes and ears,' *Modern Mechanics and Inventions,* August 1930, 168-171; '100 Trade Leaders Tour Laboratories,' *New York Times,* 8 October 1930; 'Radio-Images are Clarified,' *New York Times,* 11 January 1931; Alfred Goldsmith, 'Electrical Entertainment: A Glimpse of the Future,' *New York Times,* 22 March 1931; Earle Miller, 'What's Keeping Television Out of Your Home?' *Modern Mechanics,* May 1931, 98-102, 198, 200; 'Two-Way Television Improvements Are Sought,' *Popular Mechanics,* May 1931, 785.

> The attendant will arrange for your television conversation. When directed, please enter the booth, close the door, take your seat and turn to face the illuminated sign 'Iconophone [sic].' In a moment this sign will disappear and you will see the other party. You may then begin the conversation, a microphone and loud speaking telephone are hidden behind the drapery. Please limit your conversation to two minutes. On conclusion, turn the chair entirely around to the left and open the door.[28]

Like Hoover's appeal to the telegraph and telephone, an article entitled 'What Hath God Wrought', likened the Ikonophone conversation to the first telegraph transmission, stating more eloquently than most the steps in the process of transmitting one's face, along with the voice, over the telephone lines.

> Courtesy, personified by two youthful magicians, efficiently attending an Aladdin's lamp of complete modernity, opens the door to a sound-proof booth and bids the visitor seat himself in an upholstered swivel chair whose path of freedom is ninety degrees. The visitors swing comfortably to the right. He gazes into a black cavity at the further end of which he sees the insignia of the Bell Telephone Laboratories. There is a pause, a slight disturbance that might be called a sound.... The Bell Telephone Laboratories insigne has vanished and in its place large photograph size, is the clearly defined face of the visitor's vis-à-vis at the other end of two miles of city streets and buildings – a speaking likeness, for it has scarcely appeared on the screen before, in tone, accents and values unmistakable, come through the words: 'Why John Henry!'[29]

The reporter's lyrical tone matches the television experience well, as referring to a 'black cavity' meets the reader with a much clearer impression of what it might have been like to look through the lens. The reporter also emphasizes the astonishment felt at connecting with the remote station, illustrated in supplemental material disseminated to promote the demonstrations. At the same time, the reporter goes on to explain how he struggled to 'merely think of something not utterly banal to say.... "Did you have any trouble getting downtown?"' The novel uses for two-way television that the executives had

28 'Television Demonstration Invitation Card,' Bell Telephone Laboratories, February 1931.

29 'What Hath God Wrought!' Newspaper clipping, 18 February 1931. Container 4, Frederic Eugene Ives and Herbert Eugene Ives Papers, Manuscript Division, Library of Congress, Washington, D.C.

Figure 55. 1930 version of the Bell Labs two-way television. Interior of Television Booth. *Popular Mechanics*, June 1930.

promoted, such as establishing a more intimate connection with a loved one at a distance than the telephone could afford, was lost in banal conversations such as this. The cheeky reference in the article's title, along with the details it offers of the 'two miles of streets' between and the exclamation, reveals how the nineteenth-century hyperbolic space-annihilating rhetoric performed the duty of lifting the Machine-Age television above such otherwise forgettable communiques.

As the invitation card had promised, visitors were welcomed into a reception room at the Bell Labs office and ushered into a sound- and light-proof booth. Figure 55 provides a view of the booth's interior. This picture, widely disseminated throughout the press and in promotional material, shows a photograph in which the reporter's so-called 'black cavity' had been, perhaps in an effort to make the 'screen' more recognizable. The bright spot above the image represents the spotlight situated to illuminate the face.

The panels on either side of the display function as what one might think of as a camera. Just as the mechanical-optical approach to television replaces the screen with a neon lamp, spinning disc, and focussing lens, photoelectric cells provide the transmitting function. Behind these frosted panels hide five giant photocells, installed horizontally on the wall -- two vertically on either side and one horizontally above -- in order to detect the light bouncing off the face.

The 1930 model integrated both receiver and transmitter in each booth, which led to a host of engineering challenges. The photoelectric cells required a large amount of light in order to register an image; if the lights were too bright, they would blind the human subject and make the incoming picture impossible to see. The solution came from a combination of mechanical, electrical, and optical expertise, made possible by the highly collaborative setting of the Bell Labs research division. Ives recounted that over 100 technicians had contributed to the project, many from outside his division.[30] Ives had taken advantage of the interdisciplinarity of illuminating engineering.

The solution required a careful balance of brightness and illumination. Ives applied his expertise in light and colour measurement to the task. They landed on the solution of engineering the photocells to be sensitive to the blue range of the spectrum, which appears less harsh to the human eyes. An overhead source illuminated the small space in an orange tint.[31] Journalists drew particular attention to the absence of glaring light and the peculiarity of the blue and orange colour mixture in their articles.[32] The uniformity with which the responses appear in the press indicates that the journalists may have simply relayed the Bell Labs representative's promotional pitch.

From a science and engineering perspective, the Ikonophone had been designed according to the principles of illuminating engineering with a meticulous eye for interaction design, a practice that had its roots in human engineering and ergonomics.[33] While the particular overlaps between

30 'Television—A Group Achievement,' *Bell Laboratory Record* (May 1927): 316-317.

31 *Two-Way Television and A Pictorial account of its background* (AT&T/Bell Laboratories, 1930), 25-29; Ives, Herbert E., Frank Gray, and M. W. Baldwin, 'Image Transmission System for Two-Way Television,' *Bell System Technical Journal* 9, no. 3 (June 1930): 453.

32 'Radio's Flickering "Eyes" Now Sensitive to Color,' *New York Times,* 7 July 1929. See also '2-Way Television in Phoning Tested,' *New York Times,* 10 April 1930; 'Radio-Images are Clarified,' *New York Times,* 11 January 1931.

33 Four Bell Labs AT&T booklets present their official research and papers compiled from the Bell Labs Record and the Bell System Technical Journal. *Television,* reprinted from *Bell Laboratories Record,* June 1927; *Television: An Achievement in Electrical Communication* (American Telephone and Telegraph Company, Bell Telephone Laboratories, November 1927), consisting of a reprint of 5 papers originally presented at a Bell Labs Symposium of Television (April 1927) and published in Bell System Technical Journal (October 1927); John Mills, *Through Electrical Eyes: An elementary exposition of the physics and chemistry involved in television* (New York: Bell Telephone Laboratories, March 1928); *Two-Way Television and A Pictorial account of its background* (AT&T/Bell Laboratories, 1930). See also Herbert Ives, 'Television: 20th Anniversary,' *Bell Laboratories Record* 25, no. 5 (May 1947): 190-193; 'Birth of Television: TV Marks its Birthday,' *The Reporter,* April 1957, 4-6, 14.

illuminating engineering and human engineering come across implicitly in the literature, distinct similarities between the design philosophy and methodology suggest that they could be understood as sister disciplines. Illuminating engineering handbooks instructed that interior environments should be designed so as to make them seem as natural as possible.[34] They suggested strategies such as diffusing light sources, concealing bulbs in strategic locations, reducing glare, and tinting shade covers.

Applied to the Ikonophone booth, interaction design informed the decision to hide the telephone headset from view, concealing the presence of the human operator, and encasing the entire mechanical apparatus behind frosted glass, the colours of which were carefully chosen to create an ideal lighting environment. While they took care to account for variability in the height of the user, they removed the headset from the equation to ensure that it would not obstruct the view. The instruction card provided to visitors explains the process as if it were automatic, as if to say, 'pay no attention to the man behind the curtain' -- as in *The Wizard of Oz*, little attention is paid to the presence of the operator (Figure 42). Recounting his impression in the booth, a *New York Times* reporter wrote: 'Then the sign lifts like a magic curtain and in its place the animated picture appears of the person at the other terminal.'[35] Most visitors reporting on their experience in the booth were left with the impression that the whole process 'seems like magic'.[36]

Additionally, similar attention is paid to the chair as a kind of input mechanism. One article mentions explicitly that it only rotates 90 degrees. The instruction card, as well, seems to indicate that the swivel chair bolted to the floor performs an important role in the operation, as the instructions are explicit about users seating themselves, and standing up again at the end of the call. One writer went so far as to provide the detail that 'the act of getting up from the chair' interrupts the signal and ends the call.[37] All of these factors were meant to add up to an impression of effortlessness. As a result, the design fostered a sense that the window offered direct access to the person on the opposite end. With the goal of constructing a 'degree of realism' and an 'illusion of distance', a sense of closeness was created

34 Peter Boyce, *Human Factors in Lighting* (Boca Raton, FL: CRC Press, 2014); Stephen Guastello, *Human Factors Engineering and Ergonomics: A Systems Approach* (Boca Raton, FL: CRC Press, 2013); Christopher Nemeth, *Human Factors Methods for Design: Making Systems Human-centered* (Boca Raton, FL: CRC press, 2004). Modern textbooks on ergonomics and human-factors engineering integrate illuminating engineering and interaction design into their instruction and practices.

35 '2-Way Television in Phoning Tested,' *New York Times, 10* April 1930.

36 'Television Now Reality; Device Demonstrated,' *Troy Record,* 8 April 1927.

37 Alfred Dinsdale, *First Principles of Television* (New York: Chapman and Hall, 1932), 200.

between the two parties speaking over a distance of miles to promote the sense that they were occupying the same room.[38] The ultimate goal was to hide the fact that there was any technology involved at all.

While the space-annihilating rhetoric persisted throughout the 1920s television craze, exemplified in the way journalists wielded language like the author of 'What Hath God Wrought', it fell into the background to make way for an increased emphasis on the mastery of science as a way to control both nature and the human.[39] One popular-science monograph on television, for example, appealed to the 'annihilation of space', while noting how it would actually be accomplished through the power of science: '[Television] gives the feeling that space has been annihilated, that science to a large extent has vanquished such a thing as separation.[40] With two-way television, he then adds, the sense of closeness is far superior. But, distinct from the nineteenth-century promotion of technological progress and mastery of the inventor, the task now rested in the hands of scientists who would be able to control the viewer's sense of distance.

Defining Television

The late 1920s television craze generated a conversation about the meaning, character, and identity of a new medium. Given the variety of opinions about what television could or should do, the descriptions of its technical and visual nature ran the gamut from purely scientific to utterly science-fictional. The highly anticipated breakthrough accomplished in *actually making television work*, a technical feat that had seemed insurmountable for so long, fuelled speculation about what else could be possible with these new powers of science. Several writers remarked that, since television was now possible, it should also be possible to relay the other senses. British science writer Sydney Moseley mused that, since brain waves are just electrical impulses like the currents that pass through the cables, soon there could also be mind-reading machines. He went on:

38 Alfred Dinsdale, *First Principles of Television*, 200; *Two-Way Television and a Pictorial Account of its Background* (AT&T/Bell Laboratories, 1930), 39.

39 'What Hath God Wrought!' Newspaper clipping, 18 February 1931. Container 4, Frederic Eugene Ives and Herbert Eugene Ives Papers, Manuscript Division, Library of Congress, Washington, D.C.

40 Thomas Baker, *Wireless Pictures and Television*, 167.

> We cannot, it is true, project our bodies through space to the antipodes, but we can and do project our voices, and we can and have projected our images, so that we can be seen and heard over these vast distances, although, as yet, we cannot be felt[...][41]

Another hypothesized that 'there is no reason why we may not trick the remaining senses'.[42] The success of television broke down the barriers of credulity, making it seem like anything, including telepathy, was possible.

Breaking such barriers— like flight, space travel, and harnessing nuclear energy —has an effect on one's sense of reality.[43] Scientific and technological achievements baffle the mind, forcing us to reassess the limits of the possible. Speculative-era television discussions gravitated towards the problem of overcoming the obstacle of electrical and mechanical (technical) limitations. However, the successful demonstration of television raised awareness of the perceptual and psychological limitations standing in the way of not only making it work but making it seem real. In the words of AT&T President Walter Gifford, their two-way television was the 'fruition of years of study in the problem of seeing at a distance *as though face to face*'.[44] His words intimate that Machine-Age television had surpassed not only the nineteenth-century rhetoric of space-annihilation, but also the oversimplified confidence in technological power and progress in which it was situated. The progressive efficiency movement at the time fostered a sense that technology could facilitate the making of a better human being, a better way of organizing society. Seeing by electricity had encouraged a way of thinking about distance as a limitation that could be simply eradicated by the power of technology. Television appeared almost like portals, windows through which one maintained closeness with friends and family far away. 'Seeing at a distance *as though face to face*' introduces the notion of technological mediation, an important factor in maintaining *the perceptual illusion of closeness.*

41 Sydney Moseley, *Television Today and Tomorrow* (London: Isaac Pitman, 1929).

42 Raymond Francis Yates, *ABC of Television; Or, Seeing by Radio* (New York: Norman W. Henley Publishing, 1929), 8.

43 Michael Marshall, '10 Impossibilities Conquered by Science,' *New Scientist,* 3 April, 2008; Leo Marx, 'The Idea of "Technology" and Postmodern Pessimism,' in *Does Technology Drive History?: The Dilemma of Technological Determinism*, eds. Merritt Roe Smith and Leo Marx (Cambridge, MA: MIT Press, 1994), 237-258; Mary Ann Doane, 'Information, Crisis, Catastrophe,' in *Logics of Television: Essays in Cultural Criticism* (1990): 222-239.

44 *Two-Way Television and a Pictorial Account of its Background* (AT&T/Bell Laboratories, 1930), 4, my italics.

According to the impression relayed by one *New York Times* correspondent, the Bell Labs engineers' efforts had not been in vain:

> One has the feeling that the voice of the person at the other end of the link is issuing from the lips of the image [....] One who has experienced the sensations of the two way television–telephone booth at the laboratory feels the person at the other end of the circuit has been met face-to-face.[45]

He was not alone in this reaction. Many witnesses described the feeling of closeness established with the person on the other end. But that response, as some were keen to point out, tended to occur only when the two people knew each other well enough to recognize each other's picture. Otherwise, reactions veered towards the 'uncanny', as in referring to the image on the screen as a 'spook'.[46] Perhaps the demonstration organizers expected such a response, indicated in their suggestion that visitors invite a companion (Figure 54): 'Arrangements will be made so that you may enjoy sharing the demonstration with some associate or acquaintance to be named by you.'

Actually seeing images flash on the screen prompted the beginning of a new discussion about what television would become. In the process, scientists worked towards defining the medium by taking into account both its technical qualities and the perceptual, emotive responses it elicited. Departing from the traditional 'seeing by electricity' and 'distant electric vision', writers of popular science came up with increasingly creative definitions of the television medium:

- 'the transmission of human sight'[47]
- 'vision by wire'[48]
- instantaneous vision over any distance by wireless or wire.[49]

The inaugural issue of the short-lived British magazine *Television*, in March 1928, described television as: 'reproduction of sight, for television enables us to actually, visually, witness living scenes, people, and objects at a distance just as if we were actual eyewitnesses on the spot'.[50] In addition to appealing

45 'Radio-Images are Clarified.'

46 '100 Trade Leaders Tour Laboratories,' *New York Times*, 8 October 1930.

47 Alfred Lane, 'The Real Facts About Television,' *Popular Science*, September 1928, 43-44.

48 C. F. Jenkins, 'Life Size Radio Movies are coming,' *Modern Mechanics and Inventions*, May 1930, 70.

49 'Television 1873-1927: A brief outline of what has been accomplished in little over a century,' *Television*, March 1928, 10-11.

50 'Television 1873-1927.'

to television as a technological adjunct to sight, writers compared it to cinema. Describing television as a 'living image [...] reproduced electrically', in the case of *Science & Invention* writer Secor, suggested a comparison between the photographic film and the electrical television picture.[51] Yates, another popular-science writer, described the television medium as 'the art of *seeing* living scenes', presumably in contrast to the art of *reproducing* living scenes in the cinema.[52] His use of *seeing* as opposed to Secor's *reproducing* suggests something more complex in television's ephemerality than the transmission of light. Since scientists preferred to speak of television as the transmission of light (as opposed to images), it engendered a way of thinking about the medium as if it were as natural as human vision. Yates intimates the definition of the television image offered by Jenkins: 'There is no picture except what your eyes and brain form.'[53] It suggests that the television image is not only impossible to 'record', but that the use of 'image' only functioned in the figurative sense. The definition of television came down to a problem greater than that of medium specificity. It opened up a larger question of time, space, subjective experience, and visual perception.

In these early days of television research, engineers paid less attention to the reproduction of images than they did to the detection and reproduction of light. As the mechanics involved demanded that the visible image of a scene be broken up into millions of individual picture units (pixels), engineers more commonly referred to the 'the electrical transmission of light' as opposed to images.[54] 'We don't send pictures at all; only tiny little pieces of pictures one after the other.'[55] Ives described his approach to solving the 'problem of television', in a language characteristic of a physicist who specialized in optics, as 'the investigation of methods for producing currents in response to light variations'.[56]

Technicians tended to place emphasis on the transmitting and receiving apparatuses rather than on the abstract signal itself. This strategy encouraged observers to respond to the material apparatus before them rather than conceptualize a signal passing between stations along an electrical

51 H. W. Secor, 'Television Perfected At Last,' *Science and Invention,* June 1927, 108.

52 Yates, *ABC of Television,* 15.

53 C. F. Jenkins, 'Life Size Radio Movies are coming,' *Modern Mechanics and Inventions,* May 1930, 72.

54 'Science's Latest Marvel,' *Indianapolis Star,* 9 April 1927.

55 George H. Waltz, 'Get it on Television,' *Popular Science Monthly,* July 1931, 16-17, 136.

56 'Some points in the research and development leading to television,' in *Television: An Achievement in Electrical Communication* (American Telephone and Telegraph Company, Bell Telephone Laboratories, November 1927), 12.

wire. They filled their descriptions with references to Nipkow discs and photocells, Kerr cells and synchronizing mechanisms, which imbued the apparatus with a distinctly mechanical feel: 'In the electrical *transmission* of pictures, the transmitter is responsive to light waves [...] For the *reception* of pictures, current variations are translated into variations of light intensity.'[57]

As much as industry leaders such as Sarnoff (radio) and Alexanderson (tele-cinema) tried to impress their own views upon readers of popular-science periodicals, the Machine-Age television experiments, particularly the mechanical-optical systems of Ives, Jenkins, and Baird, constituted an approach to moving-image technology that was not only entirely novel, but also exceedingly difficult to explain to a non-technical audience. While Sarnoff was satisfied with a definition of radiovision, television as a visual adjunct to radio, and Alexanderson was content with the explanation of tele-cinema as presentation of live broadcasts in a movie theatre setting, the conceptions of television promoted by advocates of the mechanical-optical systems fell so far outside the realm of established media technologies that it practically involved the development of a new language. Even though Ives had developed two-way television as an adjunct to the telephone and Jenkins promoted radiovision as a live broadcast, the technical aspects involved in the mechanical systems overrode their attempts to explain television as an adjunct to another medium. Ives and Jenkins spoke of television as a completely new medium.

In mechanical television systems, the image only existed insofar as an observer was there to watch the spinning disc: 'light flashes into your eye', as Waltz explained. Bell's two-way television pamphlet described it as 'moving spots of varying brightness which paint a picture on the retina of the eye'.[58] Writers of popular science explained viewing the television image as a highly subjective experience. Like the falling tree that makes no sound without someone to hear it, the televised image did not exist if there was no one there to look at it: 'We really don't send pictures at all; only tiny little pieces of pictures one after the other. All the scanning disk does is to break up the picture into these tiny pieces so we can broadcast them.'[59] Secor, of *Science and Invention* magazine, wrote of 'electrical impulses representing the face'.[60]

57 *Television: An Achievement in Electrical Communication* (American Telephone and Telegraph Company, Bell Telephone Laboratories, November 1927), 1-2.
58 *First Demonstrations*, my italics. For a discussion of the optical aspects of television, see 'Image Transmission System'; Dunlap, *Outlook*, 9; Harold Horton Sheldon and Edgar Norman Grisewood, *Television: Present Methods of Picture Transmission* (New York: Van Nostrand, 1929), 21.
59 Waltz, 'Get it on Television,' 136.
60 H. W. Secor, 'Television Perfected At Last,' *Science and Invention*, June 1927, 108.

Journalists responded slightly differently. Those without any prior knowledge of electrical engineering or optics were able to respond only to their impression of the image and the interface. These accounts present attempts to reconcile experience with the technical descriptions offered. *Popular Science* tried to humanize the process, explaining that '[t]he face is divided into tiny patches of varying light and shade'.[61] The *New York Times* wrote: 'The thing that staggers the mind is that all that travelled over the wire [...] is a series of electrical impulses', adding that the person being 'televised' doesn't realize their face is being scanned 18 times a second.[62] Instead of referring to the screen or the viewer's subjective perception, *Popular Mechanics* contributor Miller chose to describe the effects of the neon lamp: 'the light source is the picture itself'.[63]

As the conversation worked to formulate a definition of television as a medium, writers also struggled to explain the abstract concept of the television signal. According to the Oxford English Dictionary, 'video' first appeared in print in 1935, in a letter to the editor of *Wireless World*.[64]

> Regarding my adjective 'visio,' I am quite ready to withdraw it provided it is agreed to use the words 'sound' and 'vision' to distinguish between the channels, tuning circuits, transmitters, and so on, connected to the two components, respectively, of a sound-and-vision programme. But I have noticed that the Americans were beginning to take 'audio' away from its original use in conjunction with 'frequency' and to use it for this special purpose; and that they were toying with the idea of 'video' as its complement. And I thought that if we were to have a fairly dreadful new word it might as will be 'visio,' which at least has the merit of being obviously connected with 'vision' in a world where compulsory Latin is rapidly dying out. And, anyhow, nobody would dare *say* 'video'; for with the vague terror of 'modern pronunciation' hanging over him he would never be sure how to pronounce it.[65]

Like the debates over choosing a proper word for 'television' as a medium and the controversies about agreeing upon a useful designation for 'looker-in',

61 George Lee Down, Jr., 'Next We'll See to Paris,' *Popular Science*, September 1927, 23.
62 'Far-off Speakers Seen as Well as Heard Here in a Test of Television: Like a Photo Come to Life,' *New York Times*, 8 April 1927; 'Two Way Television in Phoning Tested,' *New York Times*, 10 April 1930; 'Speakers on Phone See Images of Each Other,' *New York Times*, 13 April 1930.
63 James Miller, 'The Latest in Television,' *Popular Mechanics*, September 1929, 472-474.
64 'video, n.', OED Online, March 2016, Oxford University Press.
65 D'Orsay Bell, 'Alternatives to Home Television,' *Wireless World* 36, no. 3 (1935), 72;

the term 'video' ruffled feathers. To say that it did not catch on immediately would be an understatement. The modern conception of video emerged much later. In the 1950s, new methods of image processing, including coding, compression, analog-to-digital conversion, and pulse code modulation, constructed 'video' as an image signal or stream of information.[66]

Video (before 'video') emerged as a marriage of electricity, optics, and visual perception. The late-1920s conception of video took two forms. A technical explanation treated it as a transmission of light, while popular-science periodicals explained video as a technological equivalent of seeing. Neither meaning tended to use 'image' in a literal sense of the word. Rather, it encouraged a way of thinking about video as a visual perception that takes place in the brain, a cognitive and ephemeral process. The Machine-Age conception of video (before 'video') concentrated on the flickering light that forms an 'image' only when perceived by the eyes and brain. It aligned more closely with a Helmholtzian mental 'representation', a picture of the world as it exists in the mind, than with an electronic reproduction of visual images.[67] This conception contrasts sharply with the notion of photographic or cinematic representation. It also differs radically from the modern use of 'video' as an image stream processed at 30 frames per second. This earlier understanding of video as the transmission of light aligns with a particular Machine-Age construction of vision, technologically mediated.

The Television Experience

A real appreciation for the Ikonophone experience requires a separation from almost every assumption we hold about television. Mechanical television systems differ so radically from modern electronic television that it can only be resurrected today in the garages of amateur enthusiasts, or in museums devoted to dead media.[68] While it may be possible to reconstruct a working

66 C. Chapin Cutler, an oral history conducted in 1993 by Andrew Goldstein (IEEE History Center, Hoboken, NJ) 4.21-4.22; William Schreiber, an oral history conducted in 1998 by Frederik Nebeker (IEEE History Center, Hoboken, NJ) 4.4; John Pierce, an oral history conducted in 1992 by Andy Goldstein (IEEE History Center, Hoboken, NJ), Part 2, 3.9; Robert Lucky, an oral history conducted in 1999 by David Hochfelder (IEEE History Center, Hoboken, NJ), 4.5.

67 Helmholtz's theory of visual perception drew a distinction between the representation, a visual perception in the brain, and the sensation, the physical act of seeing performed by the eye. David Cahan, *Hermann Von Helmholtz and the Foundations of Nineteenth-century Science* (Berkeley, CA: University of California Press, 1993), 117-118.

68 Early Television Museum (Hilliard, OH); Danish Television historian and video producer Jan Bertelsen displays several demonstration videos on his website. See: 'Min Televisor (My

mechanical television, reconnecting to the immediate perception and the cultural context are less easily accomplished. A complete understanding of what it might have been like to look into the Ikonophone's lens must be a task we relegate to the historian's imagination.

The nineteenth-century conception of seeing by electricity continued to play a role in constructing the experience and identity of what became known as the 'looker in', the Machine-Age television viewer. The rhetoric of annihilating space prompted users to expect a direct window into a distant scene.[69] But entering the Ikonophone booth, witnesses struggled to put into words what they saw flashing in their eyes. Visitors complained about the smallness of the window, which was only about the size of a postcard.[70] Another described it as a cabinet.[71] It gave the impression of being enclosed in a sensory deprivation chamber, a far cry from the promised magic mirror.

The limited records that exist to document the conversations that occurred in the booth during these demonstrations reveal how banal the communications must have been. Several such reports were published in the press as a means to satirize and critique the two-way television. The tone of these conversations makes one wonder if two-way television was a technology anyone even wanted to have around, much less in the home or office. Stories such as these cut through the hype of the television craze, revealing the sarcastic opinions of those who expected little and feared the worst from new technologies.

These conversations suggest the dramatic difference between the promotional rhetoric and the unappealing practical applications of the two-way television. While reporters and journalists generally hesitated to suggest any practical applications of the Ikonophone, speculations of two-way television surveillance and disturbances filled lifestyle and opinion columns. Figure 33, for example, shows the two-way television invading a young woman's domestic privacy. The resurgence of a backlash against two-way television recalls the satires and critical discourses that drove the reactions against the telephonoscope and the 'Far-Sight machine'. These same reactions

Televisor),' in *And now we see by wireless* (2008); Don McLean's material on Baird's Phonovisor also proves instructional. See: tvdawn.com; Don McLean, *Restoring Baird's Image* (London: IEE, 2000).

69 'Television—At Last!' *Popular Science Monthly,* June 1927, 11-13, 130.

70 'We Catch a Glimpse of Tomorrow,' *New York Times,* 16 January 1928; 'Television Now Reality; Device Demonstrated,' *Troy Record,* 8 April 1927.

71 Alden Armagnac, 'Television Brought into the Home,' *Popular Science Monthly,* April 1928, 20-21, 143; T. R. Kennedy, 'Speakers on Phone See Images of Each Other,' *New York Times,* 13 April 1930.

returned again when Bell brought the two-way television project back in 1968 in the form of the Picturephone.[72]

Erik Barnouw, a historian and communications scholar, related his experience of attending the demonstration. He recalls his unimpressed reaction to the experience, along with the short, pointless conversation he had with his father over the Ikonophone. 'The picture was so poor, I didn't think it was worth anything at all.'[73] On a similar note, a fictionalized conversation published in the *Boston Globe* dramatized how such a conversation could go wrong, distorting the telephoner's face and connecting wrong numbers, leading the characters to conclude that, if they want to be able to see each other at a distance, they should send photos in the mail.[74] Critical responses such as these raised questions as to the value and, more importantly, the social costs of seeing by electricity.

While many of the satirical responses to the Ikonophone targeted prudish, uptight, Victorian-minded individuals cast as unwilling 'to be contaminated by the all-seeing eyes',[75] undercurrents of surveillance returned, familiar from the criticism mounted against Edison's 'Far-sight machine'. Fears that the two-way television would break down the walls of domestic privacy, particularly in the bathroom, came through with the strongest force. Opinion columns in magazines and satirical cartoons in newspapers together contributed to a sense that perhaps the two-way television was not a technology that Americans were actually likely to welcome into their homes. An opinion column in *Life* magazine dramatized the conversation between two flappers, 'quivering in agitation over the odd television contrivance'.[76] Their gossip casts television as 'perverted', 'poisonous', 'obnoxious', and 'compromising'. Another *Life* columnist proclaimed that the coming of television would require an attitude adjustment. The inevitable technological progress would mean that 'once more a Victorian inhibition will have to be swept away by the March of progress'.

The notion of the television viewer as a 'looker in' supported the popular understanding of the television screen as a window through which one could

72 Gould. See also Edward Goldstein, 'First Hand: The End of Picturephone,' EEE Oral Histories (2015).

73 Kisseloff, 21.

74 H. I. Phillips, 'The Once Over: The Television Telephone,' *Boston Globe*, 19 April 1930.

75 Edward Van Zile, 'New Perils of Television,' *The Spur*, 1 May 1928, 78.

76 Lloyd Mayer, 'Just Between Us Girls,' *Life*, 5 May 1927, 14. See also Phyllis Ryan, 'Television,' *Life*, 5 May 1927, 18.

see a distant place.[77] Journalists responding to the immediacy and ephemerality of television described the experience as if they were looking into a surreal mirror, an electric window.[78] A *Modern Mechanics* correspondent phrased it as if there was no technological mediation at all between the two speakers at either end of the Ikonophone, which was 'allowing two people to look each other in the eyes as they talk'.[79] He continued: 'Even though the speakers were situated 3 miles apart, it would not have made a particle of difference, the engineers stated, if they had been located in entire continent apart.' Reactions such as these make it seem like the nineteenth-century dream of the annihilation of space had finally been realized. Whether embedded so deep in the language of technology or having steeped so long in the American consciousness, it had finally come true.

Witnesses sometimes responded to the interaction with a television screen as if there was no mediation involved, a response that effectively reinforced the rhetoric of technological progress. Armagnac of *Popular Science* extrapolated on the window metaphor in his descriptions of the Ikonophone (Figure 32). Having been invited to a 1929 demonstration, the journalist recounts:

> Dr. Ives peered into a telescope-like window. Through a frame scarcely larger than a postage stamp he saw the young woman, startlingly lifelike, with the color and pattern of her costume perfectly reproduced. Now she held up a ball of yarn, and its crimson hue was instantly visible in the peephole receiver. Other observers took turns at the magic window.[80]

Popular-science writer Larner appealed to the specificity of the media formats when he described television as 'transmitting *actual scenes* as

77 Alfred N. Goldsmith, 'Electrical Entertainment: A Glimpse of the Future,' *New York Times*, 22 March 1931; Orrin Dunlap, 'Television Brought Nearer the Home,' *New York Times*, 25 May 1930; Dunlap, *Outlook for Television*, 141, 171; 'Latest Television Broadcast Station,' *Everyday Science and Mechanics*, November 1931, 690-691, 721; James Miller, 'The Latest In Television,' *Popular Mechanics*, September 1929, 472-474; 'Results of $50.00 "New Word" Contest,' *Television News*, July-August 1931, 211.

78 Dinsdale refers to television as 'ephemeral' in comparison to telephotography (facsimile transmission). Alfred Dinsdale, *First Principles of Television*, 5.

79 'Television Now Gives Radio eyes and ears,' *Modern Mechanics and Inventions*, August 1930, 168-171.

80 Alden Armagnac, 'Now—Television in Natural Colors,' *Popular Science Monthly*, September 1929, 25. See also 'Talk, Hear, SEE on This Phone: Two-Way Television Is Demonstrated in Laboratory As an Engineering Stunt,' *Popular Science Monthly*, July 1930, 22, 123 (describes the screen as a window).

distinct from cinema telegraphy or phototelegraphy'.[81] Defining media by their apparently formal traits—the photographic recording and the live transmission—reveals the way that technological determinism drove commercial applications. One might juxtapose this with the cheeky remark that, if telephoners wanted to be able to see each other as they talked, they would be better off sending snapshots of themselves through the mail. From this perspective, the difference between media may have more to do with facilitating communication than with technological precision.

In direct contradiction to the perception that 'television enables us to actually, visually, witness living scenes, people, and objects at a distance just as if we were actual eyewitnesses on the spot', Dunlap offers the distinction between 'the scene itself' and 'its optical counterpart'.[82] More often than not, the aesthetic of the screen as a representation overwhelmed the impression that the participants were present 'as though face to face', a response that indicated the viewer's detachment from the feeling of intimate connection. An Albany reporter, for example, described what he saw at the 1927 demonstration as 'the likeness of the speakers[...] thrown on the large screen'. He continued with a description of the two-way television picture: 'The image on the screen was like a picture postcard in clearness and about that size, but it was *a postcard come to life*; the figure moved and when the woman spoke her voice sounded clearly from the loudspeaker.'[83] Responses such as these seem to anticipate the screen as a cinematic representation, distinct from the expectations of the screen as a window, which follow from the magic-mirror legacy of seeing by electricity. Many correspondents expressed their astonishment at seeing what appears to be a motion picture come to life.

A reporter from the popular magazine *Radio Craft* described his experience in the Ikonophone as one of confusion, unable to tell the difference between the person he knew he was talking to and the appearance of that person on the screen. He wrote, 'it appeared almost as if the animated pink-and-orange image in the aperture were actually talking'.[84] Driven by expectations of cinematic quality and realism, these viewers looked at the screen with a degree of distance, constructing in their mind an imaginary

81 Larner, 147. See also 'Television 1873-1927: A brief outline of what has been accomplished in little over a century,' *Television,* March 1928, 10-11.

82 'Television 1873-1927: A brief outline of what has been accomplished in little over a century,' *Television,* March 1928, 10; Dunlap, *Outlook for Television*, 20.

83 'Television Now Reality; Device Demonstrated,' *Troy Record,* 8 April 1927.

84 Laurence M. Cockaday, 'The Latest Developments in Television Methods,' *Radio Craft,* July 1930, 22-24.

fourth wall that broke down as soon as the onscreen subject looked back. Incredulous observers responded more often with shock than astonishment. Besides the assumptions these viewers brought with them regarding medium specificity, they were also more likely to understand the screen as a representation when presented with a person they did not recognize. In the case of Brown, writing for *Nation's Business*, his immediate reaction when presented with a stranger on the screen was one of shock: 'Good Morning, Mr. Brown,' the moving picture said to me [....] Then I realized that he was seeing me just as I was seeing him. I had forgotten this in the excitement of talking to a motion picture and having a talk back to me.'[85] As engineers only began to understand much later in the development of the videophone, the technology can only support a feeling of connection between two users when that connection had already been established in face-to-face interaction.

Despite the tendency of journalists and writers of popular science to explain television as either a window or a screen, more often than not these two impressions intermingled. The conflicted, ambiguous, and ambivalent language conveyed a sense that the television viewer had both a direct, unmediated view of the person on the other end of the line as well as a live pictorial representation flickering before them on a screen. Perhaps in an effort to describe the machine as accurately as possible, Ives concocted the phrase 'the frame through which the observer sees the image of the distant person'.[86] Sometimes they used both strategies at once, as one *New York Times* reporter put it, confusing the 'living presence' of the distant televised subject with 'the flickering image of the human face'.[87]

As a way to overcome the hurdles presented by technical explanations, popular-science periodicals adopted the strategy of the studio tour. It humanized the process of televising, ushering the reader behind the scenes in a first-person narrative account. These articles dispense with any attempt at explaining the television signal. They place the emphasis entirely on the firsthand experience of seeing the transmitter and receiver in operation. The reporter often takes a conversational tone, explaining what he saw in the sequence of lights, camera, action.

Similar to the Bell Labs's 1927 demonstration auditorium set-up, which presented the visitors with a privileged view of the apparatus, these studio

85 Art Brown, 'Television is Ready for Business,' *Nation's Business,* June 1930, 47.

86 Herbert Ives, Frank Gray, and M. W. Baldwin, 'Image Transmission System for Two-Way Television,' *Bell System Technical Journal* 9, no. 3 (June 1930): 453.

87 'We Catch a Glimpse of Tomorrow,' *New York Times,* 16 January 1928.

tours pulled away the curtain to reveal the wizard at the controls. Two such articles appearing in *Popular Science* assured readers that the technical aspects of television should provide no impediment to understanding how it works and what it can do. 'Get it on Television', a transparent attempt to market new television components to the consumer, began by describing what the reporter saw on the screen:

> I could see him smile and turn his head from side to side. Then I looked through the glass windows that separated the reception room from the studio proper and there in front of some apparatus was the man himself. I had seen my first television picture, for the small outfit we were looking at was the studio's monitor set. It was tuned to reproduce whatever was being televised in front of the big machine.'[88]

In a similar fashion, *Popular Science* correspondent Armagnac narrated his experience seeing the Alexanderson system at work with a mixture of childish delight and astonishment:

> Light flickered across the window. In it appeared a face—the moving, living face of a man in the broadcasting room adjoining. 'That's Wilkins!' someone exclaimed, even before they heard the voice of Dr. Alexanderson's young assistant on the near-by loudspeaker. It *was* Wilkins—talking, grimacing, smoking a cigarette as plainly as if you were looking at him instead of seeing his image broadcast by radio![89]

These studio tours challenged reporters to negotiate between the knowledge of the person 'over there' being televised and the picture of him or her that appeared on the screen. They discarded notions of seeing by electricity and instead adopted a strategy of understanding the receiving apparatus, or screen, as a mediating device as if they were face-to-face.

A two-page spread in *Popular Mechanics*, for example, juxtaposes the electronic camera on one side, with a viewer at the receiver on the other (Figure 56).[90] Illustrating the studio layout in the background serves to demystify the technology separating the viewer from the action onscreen.

88 George H. Waltz, Jr., 'Get it on Television,' *Popular Science*, July 1931, 16-17.

89 Alden Armagnac, 'Television Brought into the Home,' *Popular Science*, April 1928, 20; see also, George Waltz, 'Get it on television,' *Popular Science*, July 1931, 16-17, 136; 'Electronic Camera "Shoots" Television Images' *Popular Mechanics*, June 1935, 878-879.

90 'Electronic Camera "Shoots" Television Images,' *Popular Mechanics*, June 1935, 878; 'London Station to Serve Ten Million,' *Popular Mechanics*, June 1935, 879.

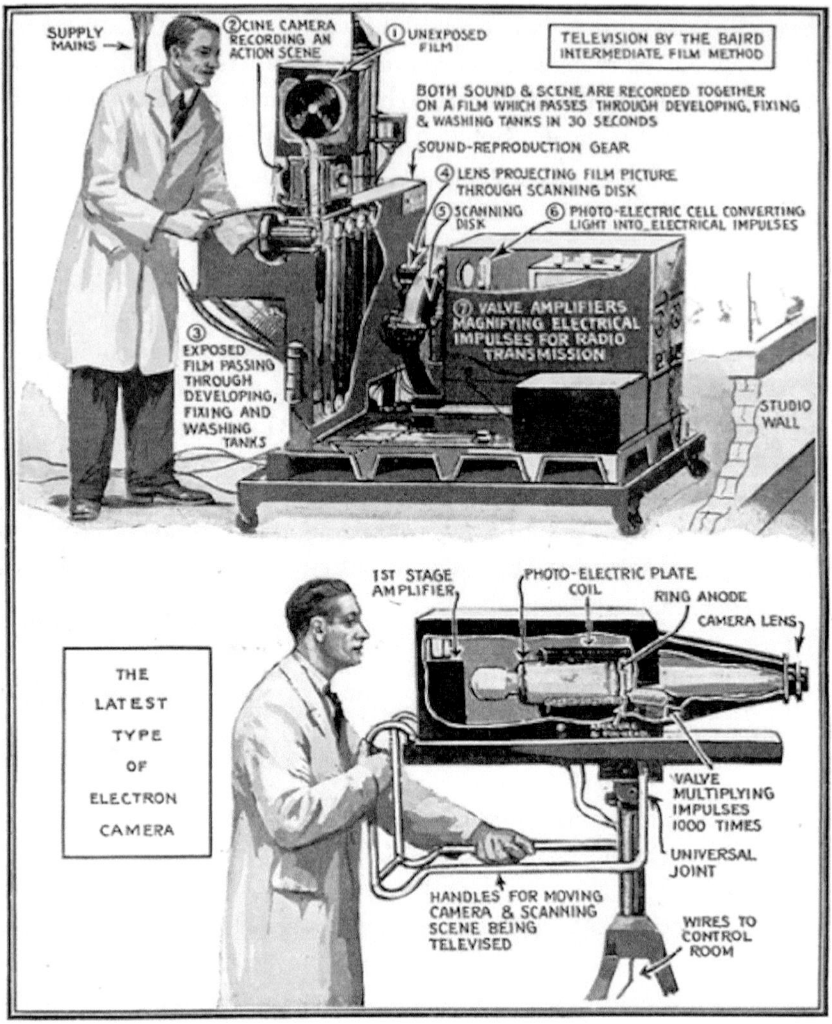

Figure 56. Opposite pages illustrating the electron camera and receiver, with a backdrop of the Baird Television Studios. 'Electronic camera "shoots" television images,' *Popular Mechanics*, June 1935, 878; 'London station to serve ten million,' *Popular Mechanics*, June 1935, 879.

While the station and home viewer, in this case, exist in remote locations, the diagram shows them connected from 'AC mains supply' and 'from aerial'. 'Get it on Television' and 'Television Brought into the home' use the same approach, illustrating the sense of spatial contiguity established in the dialogues. A collage on the first page of Waltz's article shows two overlapping pictures (Figure 57). One shows the cast posing before the camera, and below are pictured the 'prominent television engineers[...] at

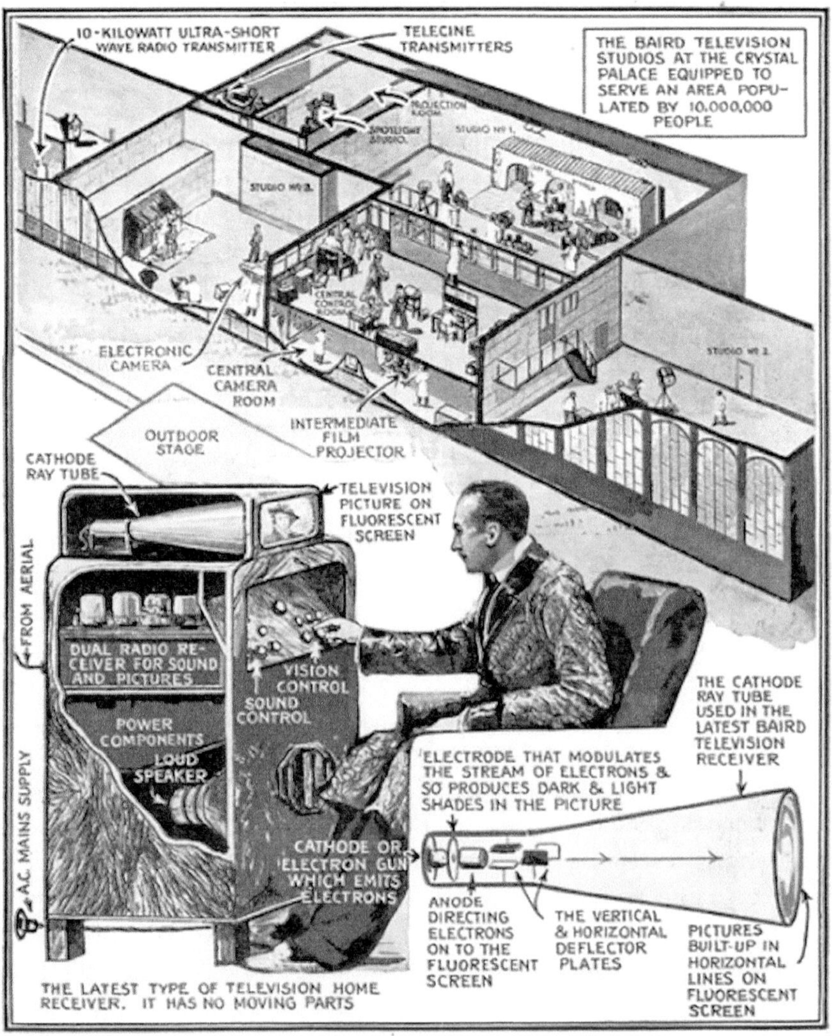

home' watching the result of their labours. Similarly, in Armagnac's article, 'Dr. Alexanderson watches the window as R. D. Kell "tunes in"'. Picturing the mechanisms driving transmitter and receiver appeals to the tinkerer who would jump at the opportunity to build a television of his own in his garage.

The rhetorical strategy worked, along with the visual depictions, to convey the sense that television was making possible a new reality of actually being able to see a distant person as if face-to-face. With television now a functional technology, no one stopped to articulate how much of a perceptual shift it would also take to get used to the difference between the recognition of physical space separating the onscreen subject and the viewer on the other end and the new construction of the television screen that was emerging to re-conceptualize space as an abstract, electronic divide. Comparing figures from Chapter Three shows how the non-technical strategy dispensed with the notion that the transmission of light would annihilate space.

These two ways of articulating the television experience (the screen as a window and as a representation) correspond with familiar approaches to theorizing the moving image (realism and illusionism). Early cinema historians consistently evoke the names of Lumière and Melies to identify the deeply rooted schism between these two approaches.[91] Gunning refers to 'the Manichean division between the films of Lumiere (documentary, realism) and the films of Melies (fiction, fantasy, stylization)', contextualizing the early formation of the binary as it developed from theatrical illusions and photographic techniques.[92]

In order to overcome this rigid binary between realism or spectacle, window or screen, film scholar Tom Elsaesser suggests levelling the historical and ontological approaches that often seem so much at odds. The old problem dividing film historians comes down to which aspect of the medium to privilege: the historical experience or its ontology. 'The question, then, is not so much: on one side spectacle, on the other narrative. Rather: we need to ask how the cinema established itself as a symbolic form.'[93] Elsaesser offers the possibility that, since neither of these approaches, spectacle nor realism, history nor ontology, fully accounts for the experience of film, a new kind of film scholarship could emerge 'from the perspective of cinema as event and experience'.[94] Accepting this compromise, the problem of screen or window would turn into a question of how the viewer interacts with the screen in both historical and ontological ways. It opens the possibility of finding relevance for 21st-century issues and trends in genealogical trajectories as well as in unchanging aspects of moving image form. The ability of such an approach to resolve the long conflict between historical and ontological film scholarship remains to be seen, for it may turn out to raise just as many new problems.

In his 2011 dissertation on the history of television, Doron Galili also touches upon the disctinction between realism and spectacle, specifically as it pertains to classical film theory. His final chapter, like this final chapter,

91 Warren Buckland, 'A Rational Reconstruction of the Cinema of Attractions,' in *Cinema of Attractions Reloaded*, 50; David Rodowick, *The Virtual Life of Film* (Cambridge, MA: Harvard University Press, 2007).

92 Gunning, 'Primitive Cinema,' 4.

93 Tom Elsaesser, 'New Film History as Media Archaeology,' 102.

94 Ibid.

Figure 57. Studio tour illustrating the onscreen subjects (above) and the engineers at home viewing their product (outline added). George Waltz, 'Get it on television,' *Popular Science*, July 1931, 16.

recognizes the transition from technology to media. When it became possible for people to see images on the screen, the identity of television changed once more from a technology to a medium. Galili writes:

> The emergence of television forced film theorists not only to speculate on its likely impact on the cinema, but also to reevaluate cinema's potential, uniqueness, and possible futures in relation to those of transmitted moving image media, which for the first time challenged film's status as the sole moving images medium. Although during that period television had not yet developed autonomous media institutions and the economic competition between the film and television had not yet begun, the recent experience of the vast changes that the coming of radio broadcasts in the 1920s brought about to mass media practices made it evident that television was about to introduce further radical shifts to the modern mediascape.[95]

The differing conceptions of moving-image technology for applications to realistic documentation and illusionistic representation come across in his moderate approach as he raises questions of both medium specificity and historical context. The late-1920s American mediascape brings these questions into relief, as the coming of sound cinema and television both contributed to the unsettling of the accepted notions of medium use based on commercial applications. Synchronous sound cinema rocked the assumptions many filmmakers had held about cinema as a primarily *visual* art form.[96] Galili's discussion of Rudolph Arnheim's film theory, contextualized in this moment of media in transition, promotes the belief that notions of media form arise from historically specific convergences of scientific, technological, and cultural forces.[97] Television demonstrations of the late 1920s presented the public with the challenge of re-conceptualizing the media landscape. Up to that point, it had existed as a visual rhetoric and discourse. But the time it spent in the physical laboratory contributed to the transformation of the technical abilities it would later be able to realize, as well as to the way it would be articulated through popular science and a new appreciation of its scientific, industrial, and technological character.

95 Galili, 183. See also Wurtzler, *Electric Sounds.*

96 Rudolph Arnheim, 'The New Laocoön: Artistic Composites and the Talking Film,' in *Film Sound: Theory and Practice,* eds. Elizabeth Weis and John Belton (McGill-Queen's Press, 1985), 112-115.

97 Galili, 212-213; Elisabeth Weis and John Belton, *Film Sound: Theory and Practice* (New York: Columbia University Press, 1985), Part Two, Section One: Classical Sound theory.

The question of medium specificity and the conflict that arose between television and cinema in the late 1920s also serves to reinforce the findings from Part I concerning the emergence of cinema. Recall in Chapter Two how the introduction of the kinetoscope drew on expectations of seeing by electricity, contributing to the expression of a paradoxical liveness to the recorded motion picture. In the 1920s, the tables turned once again. Television, now cast in the role of the newcomer, adopted some characteristics particular to the cinema. Principally, the aspects of the screen as presenting a deceptive view entered into the discourse of television. The illusionism of the cinema contributed to a way of thinking about the television screen less as a live event than as an electronic mosaic of flickering picture units that fooled the eye into thinking it was real.

The dichotomy between realism and illusion came into play in the late-1920s popular-science discourse of television. Implicitly drawing on the established conventions of the cinematic diegesis, the liveness and apparent connection established in television raised questions of the role of the screen in constructing the perception of the moving image. As an alternative to the Bell Labs engineer's turn of phrase 'as if face to face', Dinsdale referred to the 'degree of realism', describing 'holding a conversation with another person across a room at a distance of about 12 feet'.[98] In line with the 'acoustic illusion of distance' noted in the official Bell statement, Dinsdale's appeal to realism hints at the expectations that the cinema experience helped to construct. In contrast to the insistence of Bell Labs engineers on conveying a sense that two-way television offered direct access to the person on the other side, Alexanderson drew on conventions of the cinema, relying on the expectations of cinematic realism:

> Everybody lost interest in the mechanics of television; the realism was convincing enough to make them forget they were entertained by wavelengths and lenses. What the showman calls 'emotive force' came out into the theater. Those who sat before the television screen were moved by the symbols of laughter and pathos.[99]

Describing Alexanderson's long-term goals, a *Times* correspondent sketched out three perceived necessities for television's future success: picture quality, screen size, and interaction design:

98 Dinsdale, *First Principles of Television*, 200.
99 'The First Television Show,' *Popular Mechanics*, August 1930, 178.

> First, the pictures must be clear. They must be large enough so that the entire family or a theatre audience can watch the images act on a screen just as the motion–pictures actors do. And third, the television receiver that is eventually designed for home use must be as simple and fool-proof as an ordinary broadcast receiver.[100]

The goal of achieving a 'realistic' image aligned with the technological innovations driving picture quality, in which engineers and industrialists alike agreed that television would not be ready for a commercial audience until it achieved the level of realism established by the cinematic motion picture.

But those who argued that television would remain inadequate until it achieved realistic picture quality contrasted with those who revealed the deceptive nature of the electronic image. Compared to motion pictures, which present the viewer with 24 full images a second, 'in television, the eye is even more deceived'.[101] Television presents 'only a series of spots of light flashing on and off'. These writers stressed the innovative way the electronic image had been engineered, in order to draw attention to the methods involved in the scanning, deconstruction, and reconstruction of a million points of light. One Bell Labs pamphlet noted that '[t]he eye itself cannot distinguish between an object and its mirror image unless it can recognize the presence of the mirror'.[102] As one engineer put it, the expertise in telephony contributed to supporting an illusion of distance:

> The persons using the system then communicate *as if face to face* and with no telephone system *apparently* involved [....] Under these conditions, the attenuation of sounds transmitted is of about the same magnitude as would be experienced if the listener were say 10 or 12 feet away but in the same room. This acoustic illusion of distance is in harmony with the visual appearance of the television image[103]

Bell engineers simply applied the same methods already in development in their audio-research divisions to the problem of television, integrating expertise in optics and electric transmission to add two additional channels to their already robust network. With this in mind, sound studies become all the

100 Orrin Dunlap, 'Television Brought nearer the Home,' *New York Times*, 25 May 1930.

101 A. A. Albelli, 'What's Next in Television?' *Popular Mechanics*, July 1927, 3.

102 Mills, *Through electrical eyes*, 2. See also Yates, *ABC of Television*, 12: '[the television] was deceivingly human in tone and quality.'

103 *Two-Way Television and A Pictorial Account of its Vackground* (AT&T/Bell Laboratories, 1930), 37-39.

more relevant to the study of film and television history. This convergence of sound engineering with the development of moving-image technology allows us to recognize the similarities rather than the medium-specific differences between sound and moving image as *audiovisual experience.*

Popular-science periodicals imbued readers with the feeling that television was just on the horizon. Keeping their readers on the edge of their seats may have been a marketing strategy or simply a rhetorical tactic, but it also reflected the excitement and anticipation of an American audience frustrated with the tools available:

> Will tomorrow's home entertainment be furnished by a television set which, at the turn of a button, presents on a screen a visual and audible reproduction of a scene being enacted on a stage hundreds of miles away? If predictions of experimenters now working on television apparatus are to be believed, this is exactly what will be possible within a few years.[104]

Several writers promoted the view that television was possible and would reach the mainstream in a matter of years. In the late 1920s, the first generation of television broadcasters, including Jenkins (W1X1) and Cohen (W2X1, which later became CBS), encouraged radio enthusiasts to build their own sets and listen in (look in) on their evening broadcasts. Though the pictures were fuzzy and the reception sometimes spotty, these early adopters proved that television had arrived.

Additionally, some writers of popular science supported the notion that television was experiencing a unique moment in which it was still not too complicated for amateur experimenters, who could build their own receivers in a garage: 'to the legions of amateur pioneers, television falls as a rightful heritage'.[105] But for others, there was a significant difference between the technical ability to receive the television image and mastering a realistic image that any home viewer could enjoy. One *Popular Science* correspondent tried to set the story straight:

> Television, regardless of what tomorrow may bring, today is *nothing more nor less than a laboratory plaything*, a fertile field for experimentation. A

104 'Behind the Scenes with Television,' *Modern Mechanics*, January 1930, 100-101; 'Latest Television Broadcast Station,' *Everyday Science and Mechanics*, November 1931, 690-691, 721; James Millers, 'The Latest In Television,' *Popular Mechanics*, September 1929, 472-474; C. F. Jenkins, 'Life Size Radio Movies are coming,' *Modern Mechanics and Inventions*, May 1930, 70-73;
105 Yates, *ABC of Television*, 6.

The way writers articulate the meaning of new technology, the perception of mediated experience, and the so-called 'spirit of the times' continues to change along with new applications for old technologies. While definitions, applications, and possibilities of television change, the underlying human uses of technology slowly transform as well, according to updated conceptions of what it means to be human in a technologically advanced world. Such movements as cybernetics, futurism, and cyborgology attest to the burgeoning view that humans and machines exist in a symbiotic relationship.[109] The question of whether that relationship will remain in balance or shift to one side or the other remains open-ended.

Bibliography

Abramson, Albert. *The History of Television, 1880-1941.* Jefferson, NC: McFarland & Company, 1987.

Albelli, A.A. 'What's Next in Television?' *Popular Mechanics,* July 1927, 1-3.

Armagnac, Alden. 'Now—Television in Natural Colors.' *Popular Science Monthly,* Sept 1929, 25.

Armagnac, Alden. 'Television Brought into the Home.' *Popular Science Monthly,* April 1928, 20-21, 143.

Arnheim, Rudolf. 'A new Laocoön: artistic composites and the talking film.' In *Film Sound: Theory and Practice.* Edited by Elizabeth Weis and John Belton. McGill-Queen's Press, 1985: 112-115.

Baker, Thomas. *Wireless Pictures and Television.* New York : Van Nostrand, 1927.

Bell, D'Orsay. 'Alternatives to Home Television.' *Wireless World,* January 18, 1935, 72.

Boyce, Peter. *Human Factors in Lighting.* Boca Raton, FL. CRC Press, 2014.

Brown, Art. 'Television is Ready for Business.' *Nation's Business,* June 1930, 47.

Buckland, Warren. 'A Rational Reconstruction of 'The Cinema of Attractions',' in *The Cinema of Attractions Reloaded,* edited by Wanda Strauven, 41-56. Amsterdam: Amsterdam University Press, 2006.

Burns, R.W. 'The Contributions of the Bell Telephone Laboratories to the Early Development of Television.' *History of Technology* 13 (1991): 181-213.

Burns, R.W. 'Prophecy into Practice: The Early Rise of Videotelephony.' *Engineering Science & Education Journal* 4, no. 6 (1995): 33-40.

109 N. Katherine Hayles, *How We Became Post-human* (Chicago, IL: University of Chicago Press, 1999); Arthur Kroker and Marilouise Kroker, *Critical Digital Studies: A Reader* (Toronto: University of Toronto Press, 2008); Joanna Zylinska, ed. *The Cyborg Experiments: The Extensions of the Body in the Media Age* (London: A&C Black, 2002).

Cockaday, Laurence. 'The Latest Developments in Television Methods.' *Radio Craft,* July 1930, 22-4.

Coopersmith, Jonathan. *Faxed: The Rise and Fall of the Fax Machine.* Baltimore, MD: Johns Hopkins University, 2015.

Cutler, C. Chapin, an oral history conducted in 1993 by Andrew Goldstein, (IEEE History Center, Hoboken, NJ, USA).

'Demonstration of Television' invitation card (April 13, 1927), Container 3, Frederic Eugene Ives and Herbert Eugene Ives Papers, Manuscript Division, Library of Congress, Washington, D.C.

Dinsdale, Alfred. *First Principles of Television.* New York: Wiley, 1932.

Dinsdale, Alfred. *Television: Seeing by Wireless.* London: Sir I. Pitman & Sons, 1926.

Down, George Lee, Jr. 'Next We'll See to Paris.' *Popular Science,* September 1927, 23.

Dunlap, Orrin. *Outlook for Television.* New York: Harper and Brothers, 1932.

Dunlap, Orrin E. 'Television Brought Nearer the Home.' *New York Times,* May 25, 1930.

Duffus, R.L. 'The Magic of a Vast Laboratory.' *New York Times.* June 8, 1930.

Doane, Mary Ann. 'Information, Crisis, Catastrophe.' In *Logics of Television: Essays in Cultural Criticism,* edited by Patricia Mellencamp, 222-239. Bloomington: Indiana University Press, 1990.

Elsaesser Thomas. 'The New Film History as Media Archaeology.' *Cinémas: Revue d'études cinématographiques Cinémas:/Journal of Film Studies* 14, no. 2-3 (2004): 75-117.

Everyday Science and Mechanics. 'Latest Television Broadcast Station.' November 1931, 690-691, 721.

Felix, Edgar H. *Television: Its Methods and Uses.* New York: McGraw-Hill, 1931.

Friedberg, Anne. *The Virtual Window: from Alberti to Microsoft.* Cambridge, MA: MIT Press, 2006.

Galili, Doron. 'Seeing by Electricity: The Emergence of Television and the Modern Mediascape, 1878-1939.' PhD diss., University of Chicago, 2011.

Gertner, Jon. *The Idea Factory: Bell Labs and the Great Age of American Innovation.* New York: Penguin, 2012.

Goldsmith, Alfred. 'Electrical Entertainment: A Glimpse of the Future.' *New York Times.* March 22, 1931.

Goldstein, Edward. 'First Hand: The End of Picturephone.' EEE Oral Histories (2015).

Gould, John. 'Picture, Please!: How you'll be able to see while you talk on the newest version of Mr. Bell's invention.' *New York Times.* September 16, 1956.

Guastello, Stephen J. *Human Factors Engineering and Ergonomics: A Systems Approach.* Boca Raton, FL: CRC Press, 2013.

Gunning, Tom. ''Primitive' Cinema: A Frame-up? Or the Trick's on Us.' *Cinema Journal* 28, no. 2 (1989): 3-12.

Huhtamo, Erkki, and Jussi Parikka, eds. *Media Archaeology: Approaches, Applications, and Implications.* Berkeley: University of California Press, 2011.

The Independent. 'Remarkable Remarks.' October 13, 1928, 2.

Indianapolis Star. 'Science's Latest Marvel.' April 9, 1927.

Invitation card, Two-way television demonstration, 1930, Container 3, Frederic Eugene Ives and Herbert Eugene Ives Papers, Manuscript Division, Library of Congress, Washington, D.C.

Ives, Herbert. 'Television: 20th Anniversary.' *Bell Laboratories Record* 25 no. 5 (May 1947): 190-193.

Ives, Herbert E., Frank Gray, and M. W. Baldwin. 'Image Transmission System for Two-Way Television.' *Bell System Technical Journal* 9 no. 3 (June 1930): 448-469.

Jenkins, CF. 'Life Size Radio Movies are Coming.' *Modern Mechanics and Inventions*, May 1930, 70-73.

Jenkins, C. F. *Radiomovies, Radiovision, Television*. Washington, D.C.: National Capital Press, inc., 1929.

Kennedy, T.R. 'Speakers on Phone See Images of Each Other.' *New York Times*. April 13, 1930.

Kisseloff, Jeff. *The Box: An Oral History of Television, 1920-1961*. New York: Viking Press, 1995.

Lane, Alfred. 'The Real Facts About Television.' *Popular Science Monthly*, September 1928, 43-44.

Larner, Edgar T. *Practical Television*. New York: D. Van Nostrand, 1928.

Light, Jennifer. 'Facsimile: A Forgotten 'New Medium' from the 20th Century.' *New Media & Society* 8, no. 3 (2006): 355-378.

Lipartito, Kenneth. 'Picturephone and the Information Age: The Social Meaning of Failure.' *Technology and Culture* 44, no. 1 (2003): 50-81.

Lucky, Robert. an oral history conducted in 1999 by David Hochfelder, (IEEE History Center, Hoboken, NJ, USA).

Martin, Robert. 'Mystery Cell Aids Television.' *Modern Mechanics*, August 1930, 15-17.

Marx, Leo. 'The Idea of 'Technology' and Postmodern Pessimism.' In *Does Technology Drive History?: The Dilemma of Technological Determinism*, edited by Merritt Roe Smith and Leo Marx, 237-258. Cambridge, MA: MIT Press, 1994.

Mayer, Lloyd. 'Just Between Us Girls.' *Life*, May 5, 1927, 14.

McLean, Donald. *Restoring Baird's Image*. New York: Wiley-IEE, 2000.

Miller, Earle. 'What's Keeping Television Out of Your Home?' *Modern Mechanics*, May 1931, 98-102, 198, 200.

Miller, James. 'The Latest in Television.' *Popular Mechanics*, Sept 1929, 472-474.

Mills, John. *Through Electrical Eyes: An Elementary Exposition of the Physics and Chemistry Involved in Television*. New York: AT&T/Bell Telephone Laboratories, March 1928.

Mills, Mara. 'The Audiovisual Telephone: A Brief History.' In *Handheld? Music Video Aesthetics for Portable Devices*, edited by Henry Keazor, 34-47. Heidelberg, Germany: ART-Dok, 2012.

Modern Mechanics. 'Behind the Scenes with Television.' January 1930, 100-101.

Modern Mechanics. 'Television Now Gives Radio Eyes and Ears.' August 1930. 168-171.

Moseley, Sydney. *Television Today and Tomorrow.* London: Isaac Pitman, 1929.

Nemeth, Christopher. *Human Factors Methods for Design: Making Systems Human-centered.* Boca Raton, FL: CRC press, 2004.

New York Times. '100 Trade Leaders Tour Laboratories.' October 8, 1930.

New York Times. 'Deaf 'Hear' 2 Miles in Television Test.' July 3, 1930.

New York Times. 'Ethereal Images Baffle Experts.' April 20, 1930.

New York Times. 'Far-off Speakers Seen as Well as Heard Here in a Test of Television: Like a Photo Come to Life.' April 8, 1927.

New York Times. 'Radio-Images are Clarified.' January 11, 1931.

New York Times. 'Radio's Flickering 'Eyes' Now Sensitive to Color.' July 7, 1929, 124.

New York Times. 'Two-Way Television in Phoning Tested.' April 10, 1930.

New York Times. 'Washington Hails the Test: Operator There Puts Through the Calls as Scientists Watch.' April 8, 1927.

New York Times. 'We Catch a Glimpse of Tomorrow.' January 16, 1928.

Parikka, Jussi. *What is Media Archaeology.* New York: Wiley, 2013.

Phillips, H.I. 'The Once Over: The Television Telephone.' *Boston Globe.* April 19, 1930.

Pichler, Franz. 'Mechanisches Fernsehen: Fernsehempfang in der vorelektronischen Zeit' (Mechanical Television: Reception in the Pre-electronic Days). *Plus Lucis* 2 (2001): 21-26.

Pierce, John. an oral history conducted in 1992 by Andy Goldstein, (IEEE History Center, Hoboken, NJ, USA).

Popular Mechanics. 'Electronic Camera 'Shoots' Television Images.' June 1935, 878.

Popular Mechanics. 'The First Television Show.' August 1930, 177-179.

Popular Mechanics. 'London Station to Serve Ten Million.' June 1935, 879.

Popular Mechanics. 'Two-Way Television Improvements Are Sought.' May 1931, 785.

Popular Science. 'Talk, Hear, SEE on This Phone: Two-Way Television Is Demonstrated in Laboratory as an Engineering Stunt.' July 1930, 22, 123.

Popular Science. 'Television—At Last!' June 1927, 11-13, 130.

The Reporter (Bell Telephone Laboratories Magazine). 'Birth of Television: TV Marks its Birthday,' April 1957, 4-6, 14.

Rodowick, David. *The Virtual Life of Film.* Cambridge, MA: Harvard University Press, 2007.

Rubinstein, Charles. 'Optics at Bell Laboratories – General Optics, television, and Vision.' *Applied Optics* 11, no. 11 (1972): 2401-2411.

Ryan, Phyllis. 'Television.' *Life,* May 5, 1927, 18.

Schreiber, William, an oral history conducted in 1998 by Frederik Nebeker, (IEEE History Center, Hoboken, NJ, USA).

Secor, H. W. 'Television Perfected at Last.' *Science and Invention,* June 1927, 108.

Shapin, Steven and Simon Shaffer. *Leviathan and the Air-Pump: Hobbes, Boyle and the Experimental Life.* Princeton, NJ: Princeton University Press, 1985.

Sheldon, H.H. and Edgar Norman Grisewood. *Television: Present Methods of Picture Transmission.* New York: D. Van Nostrand Company, Inc., 1929.

Shiers, George. 'The Rise of Mechanical Television, 1901–1930.' *SMPTE Journal* 90, no. 6 (1981): 508-521.

Television: An Achievement in Electrical Communication. New York: AT&T/ Bell Telephone Laboratories, November 1927.

'Television.' Special Issue, *Bell Laboratory Record* 4 no. 3 (May 1927). Articles including: 'The Demonstration at Washington,' 'Television by Radio,' 'Research and Development Leading to Television,' 'Physical Principles and Apparatus,' and 'Television—A Group Achievement.'

Television News. 'Results of $50.00 'New Word' Contest.' July 1931, 211.

Troy Record. 'Television Now Reality; Device Demonstrated.' April 8, 1927.

'Two-Way Television.' *The Scientific Monthly* 30 (May 1930): 476-480.

Two-Way Television and A Pictorial account of its background. New York: AT&T/ Bell Laboratories, 1930.

'What Hath God Wrought!' Newspaper clipping, February 18, 1931. Container 4, Frederic Eugene Ives and Herbert Eugene Ives Papers, Manuscript Division, Library of Congress, Washington, D.C.

Van den Ende, Jan, Wim Ravesteijn, and Dirk De Wit. 'Shaping the Early Development of Television.' *Technology and Society Magazine, IEEE* 16, no. 4 (1997): 13-26.

Van Zile, Edward. 'New Perils of Television.' *The Spur,* May 1, 1928, 78.

Weber, Anne-Katrin. 'Recording on Film, Transmitting by Signals: The Intermediate Film System and Television's Hybridity in the Interwar Period.' *Grey Room* 56 (2014): 6-33.

Weis, Elisabeth, and John Belton. *Film Sound: Theory and Practice.* New York: Columbia University Press, 1985.

Waltz, George H. 'Get it on Television.' *Popular Science,* July 1931, 16-17, 136.

Washington Post. 'Two-Way Television Closes Two-Mile Gap.' April 9, 1930.

Wurtzler, Steve. *Electric Sounds: Technological Change and the Rise of Corporate Mass Media.* New York: Columbia University Press, 2007.

Wythoff, Grant. 'Pocket Wireless and the Shape of Media to Come, 1899–1922.' *Grey Room* 51 (2013): 40-63.

Yates, Raymond F. *ABC of Television; Or, seeing by Radio.* New York: Norman W. Henley Publishing, 1929.

Yuste, Antonio Perez, 'La televisión mecánica' (The Mechanical Television). In *Detrás de la Cámara. Historia de la Televisión y de sus Cincuenta Años en España* (*Behind the Scenes: History of Television and its Fifty Years in Spain*), 65-82. Madrid: Colegio Oficial de Ingenieros de Telecomunicación, 2008.
Zeise, Tina. *Geschichte und Technik des analogen Fernsehens* (*History and technology of Analog Television*). Munich: GRIN Verlag, 2006.

Epilogue

If you believe FaceTime and Skype are 21st-century inventions, think again. In the late nineteenth and early 20th centuries, speculations on the future possibility of 'tele-vision', alternatively called 'seeing by electricity', 'distant electric vision', or identified by a host of individual inventions (telectroscope, telephonoscope, telephote, etc.), circulated ravenously. Science-fiction authors invented on the page. So did a great many journalists, engineers, and scientists who, between 1880 and 1930, contributed to the discourse of 'television'. Television's speculative era occurred long before TV sets ever made their way into domestic living spaces.

Though, today, it is common to hear television producers and studios -- even viewers bemoan the death of the television as a mass medium -- the concept of television may turn out to be a medium in transition. Television is not necessarily a *fixed idea*. Strip away the assumptions and expectations and speculations about television, and you will find an elaborate network of communication. At this stage, we need to ask: What distinguished the emerging 21st-century media landscape from that of the late nineteenth century? Themes of access to information, fears and anxieties of surveillance, and screen mediation occur persistently throughout television's lifespan.

The visual rhetoric and discourse of speculative-era televisual culture was in the hands of a small group of journalists and authors, but the Internet offers a voice to anyone with accessibility to the network. One thing that has not significantly changed is the digital divide, indicated in Du Maurier's 'Edison's Telephonoscope' (Figure 1), between East and West. The Sri Lankan nursemaid still has no voice: no internet access, mobile telephone, nor computer.

This very term 'participatory culture' may be reflexive as well, since nineteenth century technological folklore indicates that media cultures were participatory even before they were interactive. Rumours circulated like viruses and memes. Newspaper and magazine readers wrote to editors and had their letters published, eventually contributing to a broader, evolving televisual culture.

Roberts, I., *Visions of Electric Media: Television in the Victorian and Machine Ages.*
Taylor & Francis Group, 2019
DOI 10.5117/9789462986596_EPI

[When we have Television shall we ever be given the wrong number to look at while listening to the right one?]

Voice over wire. "AND DOES HONEY-BUNCH LOVE BUNNY-BOY AS MUCH AS EVER?"

Figure 58. 'When we have Television...' *Punch*, 4 May 1927.

Secondly, surveillance resurfaces as a persistent anxiety within televisual culture. The threat of surveillance cannot be ignored today. A library of literature has amassed on 'surveillance society', offering positive and negative opinions as to the potential of CCTV, web cameras, and invasive hackers. Too many versions of Panopticism have been speculated upon.

Yet, surveillance society was an enduring aspect of television's speculative era. Edison's 'Far-Sight Machine', that all-seeing eye, threatened to shine a 'fierce white light' on society.[1] Similar fears of surveillance and disruption permeated the reception of the Ikonophone in the 1920s.[2] Like 'Face with the Smile' (Figure 33), Heath Robinson's 'Some uses for television' illustrates the ludicrous ways 'distant electric vision' could make life easier, bring the world closer, and complicate situations at the same time. The ever-reliable *Punch* offered their impression of the Ikonophone in a 1927 illustration (Figure 58), recalling criticisms lodged against the intrusive 'Far-Sight Machine' in 1889. A column published in *The Spur* mirrored the outrage expressed by those nineteenth-century satirists: 'It would be the greatest of all calamities should the latter ['its men of vision'; the clergy, politicians, poets] become

1 'Edison's Last,' *Boston Journal*, 13 May 1889.

2 Phyllis Ryan, 'Television,' *Life*, 5 May 1927, 18; Lloyd Mayer, 'Just Between us Girls,' *Life*, 5 May 1927, 14; Edward van Zile, 'New Perils of Television,' *The Spur*, 1 May 1928, 78; H. I. Phillips, 'The Once Over: On the Television Telephone,' *Daily Boston Globe*, 18 April 1930, 14.

dumb through the irreverent audacity of television.'[3] Columns published in *Life* responded to the prospect of 'two-way television' with similar outrage, though with a more sarcastic and cheeky tone. 'Just between us girls' staged a conversation between two flappers:

> I am getting rapidly ossified with this vile idea because it just means that you will have the most sobering experience like having some boy friend suddenly phone you when you are bounding around your boudoir [....] Think of the people you could call up and discover in all kind of Compromising and embarrassing positions.[4]

While the nineteenth-century criticism had located the spaces of clandestine activity in the opera house and the man's club, 20th-century criticism focussed almost exclusively on the dressing room and the bathtub, more reflective of a fear of nudity than a fear of being caught in a publicly shameful act.[5]

Satires of 'television' challenge us to consider a comparison between what might be perceived as 'unmediated' communication and the effects that the introduction of new technologies have on the perception of mediation. The critical response to the 'Far-Sight Machine' centred around the difference between what was perceived as the unmediated social interaction in the opera or theatre auditorium and what was thought to be the introduction of a technologically mediated sight presented by the electric telescope. Illustrations of the telephonoscope and the 'Far-Sight Machine' (Figures 2, 16) show how mediation was perceived by some as getting in the way of 'real', 'unmediated' social engagement. Neglecting to fasten the telephonoscope peephole makes private spaces susceptible to invasion. From this perspective, the greatest difference between the nineteenth-century culture of seeing by electricity and the electronic screens engineered by Machine-Age scientists rests on the fact that the former were criticized for introducing a technologically mediated form of vision. The latter came to be understood as a technological way of seeing.

While the nineteenth-century culture of seeing by electricity rejected the notion that technology would be able to, or should be able to, provide a direct link between correspondents, the 20th-century notions of technological devices as able to provide automatic and direct communication grew

3 Edward van Zile, 'New Perils of Television,' *The Spur*, 1 May 1928, 78.

4 Lloyd Mayer, 'Just Between us Girls,' *Life*, 5 May 1927, 14.

5 'When the Movies Fly Away' and 'New Peril' also mention bathing.

prominent as the result of a heightened emphasis on the power of science to foster human mastery over nature. Having broken the mould established by the telegraph office, which installed a human interface in the form of the telegraph clerk between the general public and the technological apparatus, the telephone introduced the notion that an electro-mechanical device could facilitate a direct link between two distant parties. The telephonoscope pushed that proposition one step further with the suggestion that the machine interface, so apparent in the telephone's bulky headset, could become transparent in the form of the mirror or screen. Related to the theme of space-annihilation driving much of the progressive rhetoric, the debate came down to the 'revolutionary' proposition that the machine interface would replace the human interface.[6]

But more troubling is the lack of historical perspective, along with the proliferation of myths about media origination. Historical context gets swept under the rug in lieu of spectacle and hyperbole. The appearance of a culture of seeing by electricity in the nineteenth century is all the more valuable to providing a basis for the study of different concepts of television because at that time the devices only existed in the mind, on paper, and in discourse. Seeing by electricity condensed all these desires and fears, anxieties and expectations blown out of proportion in the rhetoric and discourse of space annihilation. It promised a change radical enough to unify the conservative attitude, rejecting the notion that a machine could ever take the place of what had been perceived as an 'unmediated' social and technological environment. But in retrospect one must also consider how the nineteenth-century technological environment already constructed mediated relationships. Just as the function of the telegraph clerk as human interface went unnoticed, the opera box and theatre auditorium architecture, which divided its audience into a physical hierarchy, could be understood as natural in relation to the artificiality of electric opera glasses. Analysing these differences serves as a reminder that 'unmediated' communication like direct access, are illusions that serve to mystify the methods of control used in constructing all aspects of modern experience.

Even though many themes remained consistent throughout television's speculative era, several important historical changes took place. As discussed in Chapter Three, a new philosophy of technology founded in systems thinking and electronic engineering displaced the mechanical approach taken in the nineteenth century. Along the way, Machine-Age engineering philosophy shaped the new conception of what television could become,

6 'Edison's Last.'

what it ought to be. One key difference between the nineteenth- and 20th-century constructions of 'television' comes down to the rhetorical approaches involved in articulating its meaning in the press. While expressions of 'space annihilation' and the conquest of nature continued through 1930 as an important part of the rhetoric of television's development, a new tendency emerged related to the scientific and technological character of 'television'. As television grew into a visual medium, distinct from the technological apparatuses of transmitter and receiver, viewers began to recognize the ways in which the screen itself constructed their experience of using what they alternately referred to as a window and a frame. The concept of television transformed from a magic mirror thought to provide unmediated access to a distant place or person, to a notion that the screen mediated the user's face-to-face interaction.

In choosing to focus on the changes that occurred across television's speculative era, a few important things become apparent. For one thing, cultures in different periods of time receive new technologies with hysteria and hyperbole. The discovery mania of 1878, the spectacle of the cinema, the satire of the far-sight machine, and the television craze of the 1920s all mark moments in American history when a new medium seemed to be on the brink of arrival. The series of failures that came about throughout this period of speculation, however, show how expectations rub up against the reality of technological possibilities and frame the reception of new media. In each case, the seemingly 'new' technology came wearing new clothes. 'Television' adapts to the expectations thrust upon it in each new age.

Bibliography

Boston Journal. 'Edison's Last.' May 13, 1889.

Hayles, N. Katherine. *How We Became Posthuman: Virtual Bodies in Cybernetics, Literature, and Informatics.* Chicago, IL: University of Chicago Press, 1999.

Kroker, Arthur, and Marilouise Kroker, ed. *Critical Digital Studies: A Reader.* Toronto: University of Toronto Press, 2008.

Mayer, Lloyd. 'Just Between Us Girls.' *Life,* May 5, 1927, 14.

Phillips, H.I. 'The Once Over: The Television Telephone.' *Boston Globe.* April 19, 1930.

Ryan, Phyllis. 'Television.' *Life,* May 5, 1927, 18.

Van Zile, Edward. 'New Perils of Television.' *The Spur,* May 1, 1928, 78.

Zylinska, Joanna, ed. *The Cyborg Experiments: The Extensions of the Body in the Media Age.* London: A&C Black, 2002.

Bibliography

Secondary Sources

Abramson, Albert. *The History of Television, 1880-1941.* Jefferson, NC: McFarland & Company, 1987.

Alexander, Jennifer. 'Efficiencies of Balance Technical Efficiency, Popular Efficiency, and Arbitrary Standards in the Late Progressive Era USA.' *Social Studies of Science* 38, no. 3 (2008): 323-349.

Alexander, Jennifer. *The Mantra of Efficiency: From Waterwheel to Social Control.* Baltimore, MD: Johns Hopkins University Press, 2008.

Alkon, Paul K. *Science Fiction Before 1900: Imagination Discovers Technology.* Woodbridge, CT: Twayne Publishers, 1994.

Allen, Robert. 'Relocating American Film History: The "Problem" of the Empirical.' *Cultural Studies* 20, no. 1 (2006): 48-88.

Altman, Rick. *Silent Film Sound.* New York: Columbia University Press, 2007.

Altick, Richard. *Punch: The Lively Youth of a British Institution, 1841-1851.* Columbus, OH: Ohio State University Press, 1997.

Anderson, Joseph, and Barbara Fisher. 'The Myth of Persistence of Vision.' *Journal of the University Film Association* 30, no. 4 (1978): 3-8.

Anderson, Joseph and Barbara Anderson. 'The Myth of Persistence of Vision Revisited.' *Journal of Film and Video* (1993): 3-12.

Andriopoulos, Stefan. *Ghostly Apparitions: German Idealism, the Gothic Novel, and Optical Media.* New York: Zone Books, 2013.

Andriopoulos, Stefan. 'Psychic Television.' *Critical Inquiry* 31, no. 3 (2005): 618-637.

Arnheim, Rudolf. 'A new Laocoön: artistic composites and the talking film.' In *Film Sound: Theory and Practice.* Edited by Elizabeth Weis and John Belton, 112-115. New York, NY: Columbia University Press, 1985.

Arnold, Darrell, ed. *Traditions of Systems Theory: Major Figures and Contemporary Developments.* New York: Routledge, 2014.

Ashley, Michael, and Robert Lowndes. *The Gernsback Days: A Study of the Evolution of Modern Science Fiction from 1911 to 1936.* Holicong, PA: Wildside Press LLC, 2004.

Balbi, Gabriele. 'Radio Before Radio: Araldo Telefonico and the Invention of Italian Broadcasting.' *Technology and Culture* 51, no. 4 (2010): 786-808.

Barnes, John. *The Beginnings of the Cinema in England 1894-1901.* London: University of Exeter Press, 1997.

Barnouw, Erik. *Tube of Plenty: The Solution of American Television.* Cambridge, UK: Oxford University Press, 1990.

Barthes, Roland. *Image-Music-Text.* London: Fontana Press, 1977.

Batchen, Geoffrey. *Burning with Desire: The Conception of Photography.* Cambridge, MA: MIT Press, 1999.

Baxandall, Michael. *Patterns of intention: On the historical explanation of pictures.* New Haven, CT: Yale University Press, 1985.

Baym, Nancy. *Personal Connections in the Digital Age.* Malden, MA: Polity, 2010.

Bazerman, Charles. *The Languages of Edison's Light.* Cambridge, MA: MIT Press, 1999.

Bazin, André. 'The Myth of Total Cinema.' In *What is Cinema.* Translated by Hugh Gray, 17-22. Berkeley, CA: University of California Press, 1967.

Beegan, Gerry. *The Mass Image: A Social History of Photomechanical Reproduction in Victorian London.* London: Palgrave Macmillan UK, 2008.

Beniger, James. *The Control Revolution: Technological and Economic Origins of the Information Society.* Cambridge, MA: Harvard University Press, 2009.

Biersack, Aletta, and Lynn Avery Hunt. *The New Cultural History.* Berkeley, CA: University of California Press, 1989.

Bijker, Wiebe, Thomas Hughes, and Trevor Pinch, eds. *The Social Construction of Technological Systems: New Directions in The Sociology and History of Technology.* Cambridge, MA: MIT Press, 1987.

Biocca, Frank, Chad Harms, and Judee K. Burgoon. 'Toward a More Robust Theory and Measure of Social Presence: Review and Suggested Criteria.' *Presence* 12, no. 5 (2003): 456-480.

Blank, Trevor, ed. *Folk Culture in the Digital Age: The Emergent Dynamics of Human Interaction.* Boulder, CA: University Press of Colorado, 2012.

Bonnell, Victoria E., Lynn Avery Hunt, and Richard Biernacki. *Beyond the Cultural Turn: New Directions in the Study of Society and Culture.* Berkeley, CA: University of California Press, 1999.

Bottomore, Stephen, 'The Panicking Audience?: Early Cinema and the "Train Effect".' *Historical Journal of Film, Radio and Television* 19, no. 2 (1999): 177-216.

Bourdieu, Pierre. *The Field of Cultural Production: Essays on Art and Literature.* New York: Columbia University Press, 1993.

Boyce, Peter. *Human Factors in Lighting.* Boca Raton, FL: CRC Press, 2014.

Braun, Marta *Eadweard Muybridge.* London: Reaktion Books, 2012.

Braun, Marta. *Picturing Time: The Work of Etienne-Jules Marey (1830-1904).* Chicago, IL: University of Chicago Press, 1994.

Braun, Marta, Charlie Keil, Rob King, Paul Moore, and Louis Pelletier, eds. *Beyond the Screen: Institutions, Networks and Publics of Early Cinema.* New Barnet, UK: John Libbey, 2012.

Brey, Philip. 'Technology as Extension of Human Faculties.' In *Metaphysics, Epistemology and Technology.* Edited by Carl Mitcham, 59-78. London: Elsevier/ JAI Press, 2000.

Brill, Michael. 'How the CIE 1931 Color-Matching Functions were Derived from Wright-Guild Data.' *Color Research & Application* 23, no. 4 (1998): 259-259.

Brooker, Will, and Deborah Jermyn, ed. *The Audience Studies Reader.* London: Routledge, 2003.

Buckland, Warren. 'A Rational Reconstruction of "The Cinema of Attractions",' in *The Cinema of Attractions Reloaded.* Edited by Wanda Strauven, 41-56. Amsterdam: Amsterdam University Press, 2006.

Burgess, Jean, and Joshua Green. *YouTube: Online Video and Participatory Culture.* Medford, MA: Polity Press, 2013.

Burns, R. W. 'The Contributions of the Bell Telephone Laboratories to the Early Development of Television.' *History of Technology* 13 (1991): 181-213.

Burns, R. W. 'Prophecy into Practice: The Early Rise of Videotelephony.' *Engineering Science & Education Journal* 4, no. 6 (1995): 33-40.

Burns, R. W. *Television: An International History of the Formative Years.* London: Institutions of Electrical Engineers, 1998.

Butsch, Richard. *The Making of American Audiences: From Stage to Television, 1750-1990.* Cambridge, UK: Cambridge University Press, 2000.

Cahan, David. *Hermann Von Helmholtz and the Foundations of Nineteenth-century Science.* Berkeley, CA: University of California Press, 1993.

Campanella, Thomas. 'Eden by Wire: Web Cameras and the Telepresent Landscape.' In *The Robot in the Garden: Telerobotics and Telepistemology in the Age of the Internet.* Edited by Ken Goldberg, 22-47. Cambridge, MA: MIT Press, 2001.

Canales, Jimena. *A Tenth of a Second: A History.* Chicago, IL: University of Chicago Press, 2010.

Carlson, Bernard. 'Artifacts and Frames of Meaning: The Cultural Construction of Motion Pictures.' In *Shaping Technology/Building Society: Studies in Sociotechnical Change.* Edited by Wiebe E. Bijker and John Law, 175-198. Cambridge, MA: MIT Press, 1992.

Cartwright, Lisa. *Screening the Body: Tracing Medicine's Visual Culture.* Minneapolis, MN: University of Minnesota Press, 1995.

Castleman, Harry, and Walter J. Podrazik. *Watching TV: Six Decades of American Television.* Syracuse, NY: Syracuse University Press, 2003.

Checkland, Peter. *Systems Thinking, Systems Practices.* New York: John Wiley, 1981.

Carlson, Bernard, and Michael E. Gorman. 'Understanding Invention as a Cognitive Process: The Case of Thomas Edison and Early Motion Pictures, 1888-91.' *Social Studies of Science* 20, no. 3 (1990): 387-430.

Chun, Wendy. 'Introduction: Did Somebody Say New Media?' In *New Media, Old Media: A History and Theory Reader.* Edited by Wendy Chun and Thomas Keenan. New York: Routledge, 2006.

Clayton, Jay. *Charles Dickens in Cyberspace: The Afterlife of the Nineteenth Century in Postmodern Culture.* Cambridge, UK: Oxford University Press, 2003.

Cochrane, Rexmond. *Measures for Progress: A History of the National Bureau of Standards.* Washington, D. C.: US Department of Commerce, 1966.

Codell, Julie. 'Imperial Differences and Culture Clashes in Victorian Periodicals' Visuals: The Case of *Punch.*' *Victorian Periodicals Review* 39, no. 4 (2006): 410-428.

Cohn, Richard Steven. 'Who Put the Magic in Movies?' *SMPTE Motion Imaging Journal* (February/March 2006): 80-94.

Collins, Theresa, Lisa Gitelman, and Gregory Jankunis. *Thomas Edison and Modern America: A Brief History with Documents.* New York: Palgrave Macmillan, 2002.

Conot, Robert. *A Streak of Luck: The Life and Times of Thomas A. Edison.* Boston, MA: Da Capo Press, 1979.

Coopersmith, Jonathan. *Faxed: The Rise and Fall of the Fax Machine.* Baltimore, MD: Johns Hopkins University, 2015.

Corwin, Sharon. 'Picturing efficiency: Precisionism, Scientific Management, and the Effacement of Labor.' *Representations* 84, no. 1 (2003): 139-165.

Coté, Mark. 'Technics and the Human Sensorium: Rethinking Media Theory Through the Body.' *Theory & Event* 13, no. 4 (2010).

Crary, Jonathan. *Techniques of the Observer: On Vision and Modernity in the Nineteenth Century.* Cambridge, MA: MIT Press, 1992.

Croteau, David, and William Hoynes. *Media/ Society: Industries, Images, and Audiences.* Thousand Oaks, CA: Sage, 2014.

Csicsery-Ronay, Jr., Istvan. *The Seven Beauties of Science Fiction.* Middletown, CT: Wesleyan University Press, 2008.

Cubitt, Sean. *The Practice of Light: A Genealogy of Visual Technologies from Prints to Pixels.* Cambridge, MA: MIT Press, 2014.

Daly, Nicholas. 'The Machine Age.' In *The Oxford Handbook of Modernisms.* Edited by Peter Osborne, Peter Brooker, Andrzej Gasiorek, Deborah Longworth, and A. J. Thacker, 283-299. Cambridge, UK: Oxford University Press, 2010.

Danziger, Kurt. *Constructing the Subject: Historical Origins of Psychological Research.* Cambridge, UK: Cambridge University Press, 1994.

Darnton, Robert. *The Great Cat Massacre and Other Episodes in French Cultural History.* New York: Basic Books, 1984.

Darnton, Robert. 'The Symbolic Element in History.' *The Journal of Modern History* 58, no. 1 (1986): 218-234.

Darnton, Robert. 'What is the History of Books?' *Daedalus* (1982): 65-83.

Daston, Lorraine, and Peter Galison. 'The Image of Objectivity.' *Representations* (1992): 81-81.

Daston, Lorraine, and Peter Galison. *Objectivity.* Cambridge, MA: MIT Press, 2007.

Davis, Lennard J. 'Constructing Normalcy: The Bell Curve, the Novel, and the Invention of the Disabled Body in the Nineteenth Century.' In *The Disability Studies Reader*, 2nd edition. Edited by Lennard J. Davis, 3-16. New York: Taylor & Francis, 2006.

De Laurentis, Teresa, Andreas Huyssen, and Kathleen M. Woodward, eds. *The Technological Imagination: Theories and Fictions*. Madison, WI: Coda Press, 1980.

Dienst, Richard. *Still Life in Real Time: Theory After Television*. Durham, NC: Duke University Press, 1994.

Doane, Mary Ann. *The Emergence of Cinematic Time: Modernity, Contingency, the Archive*. Cambridge, MA: Harvard University Press, 2002.

Doane, Mary Ann. 'Information, Crisis, Catastrophe.' In *Logics of Television: Essays in Cultural Criticism*. Edited by Patricia Mellencamp, 222-239. Bloomington, IN: Indiana University Press, 1990.

Dorson, Richard, ed. *Folklore and Folklife*. Chicago, IL: University of Chicago Press, 1972.

Douglas, Susan J. *Inventing American Broadcasting, 1899-1922*. Baltimore, MD: Johns Hopkins University Press, 1989.

Edgerton, Gary. *The Columbia History of American Television*. New York: Columbia University Press, 2010

Elkins, James. *Visual Studies: A Skeptical Introduction*. New York: Psychology Press, 2003.

Elsaesser Thomas. 'The New Film History as Media Archaeology.' *Cinémas: Revue d'études cinématographiques Cinémas:/Journal of Film Studies* 14, no. 2-3 (2004): 75-117.

Ernst, Waltraud. 'The Normal and the Abnormal: Reflections on norms and normativity.' In *Histories of the Normal and the Abnormal: Social and Cultural Histories of Norms and Normativity*. Edited by Waltraud Ernst, 1-25. New York: Routledge, 2006.

Ernst, Wolfgang. *Digital Memory and the Archive*. Edited by Jussi Parikka. Minneapolis, MN: University of Minnesota Press, 2013.

Essig, Mark. *Edison and the Electric Chair: A Story of Light and Death*. New York: Bloomsbury Publishing USA, 2005.

Fickers, Andreas. 'Television.' In *The Handbook of Communication History*. Edited by Peter Simonson, Janice Peck, Robert T. Craig, and John Jackson, 239-255. New York: Routledge, 2013.

Finkelstein, David, and Alistair McCleery, eds. *The Book History Reader*. New York: Routledge, 2002.

Foucault, Michel. *The Archaeology of Knowledge*. New York: Pantheon, 1972.

Friedberg Anne. *The Virtual Window: from Alberti to Microsoft*. Cambridge, MA: MIT Press, 2006.

Galili, Doron. 'Seeing by Electricity: The Emergence of Television and the Modern Mediascape, 1878-1939.' PhD diss., University of Chicago, 2011.

Gaudreault, André, ed. *American Cinema, 1890-1909: Themes and Variations.* New Brunswick, NJ: Rutgers University Press, 2009.

Gaudriault, Andre. *Film and Attraction: From Kinematography to Cinema.* Chicago, IL: University of Illinois Press, 2011.

Gaycken, Oliver. 'Introduction on Displaying knowledge: Intermedial Education.' *Early Popular Visual Culture* 13, no. 4 (2015): 249-255.

Gaycken, Oliver. *Devices of Curiosity: Early Cinema and Popular Science.* Cambridge, UK: Oxford University Press, 2015.

Gaycken, Oliver. 'The School of the Future: Thomas A. Edison, Inc. and the Popular Science Film.' Lecture, Virginia Commonwealth University, Richmond, VA, 6 April 2016.

Gaycken, Oliver. '"The Swarming of Life": Moving Images, Education, and Views through the Microscope.' *Science in Context* 24.03 (2011): 361-380.

Gertner, Jon. *The Idea Factory: Bell Labs and the Great Age of American Innovation.* New York: Penguin, 2012.

Gitelman, Lisa. *Always Already New: Media, History and the Data of Culture.* Cambridge, MA: MIT Press, 2006.

Gitelman, Lisa. *Scripts, Grooves and Writing Machines: Representing Technology in the Edison Era.* Stanford, CA: Stanford University Press, 1999.

Gomery, Douglas. *A History of Broadcasting in the United States.* New York: Wiley-Blackwell, 2008;

Gooday, Graeme. *Domesticating Electricity: Technology, uncertainty and gender, 1880–1914.* London: Pickering & Chatto, 2008.

Gooday, Graeme. *The Morals of Measurement: Accuracy, Irony, and Trust in Late Victorian Electrical Practice.* Cambridge, UK: Cambridge University Press, 2004.

Grau, Oliver. 'The History of Telepresence: Automata, Illusion, and Rejecting the Body.' In *The Robot in the Garden: Telerobotics and Telepistemology in the Age of the Internet.* Edited by Ken Goldberg, 226-243. Cambridge, MA: MIT Press, 2001.

Grau, Oliver, and Thomas Veigl. *Imagery in the 21st Century.* Cambridge, MA: MIT Press, 2011.

Grau, Oliver. *Virtual Art: From Illusion to Immersion.* Cambridge, MA: MIT Press, 2003.

Gunning, Tom. 'An Aesthetic of Astonishment: Early Film and the (In)credulous Spectator.' *Art and Text* 34 (Spring 1989): 31-45.

Gunning, Tom. 'The Cinema of Attraction.' *Wide Angle* 3, no. 4 (1986).

Gunning, Tom. 'Doing for the Eye What the Phonograph Does for the Ear.' In *The Sounds of Early Cinema.* Edited by Richard Abel and Rick Altman, 13-31. Bloomington, IN: Indiana University Press, 2001.

Gunning, Tom. 'Phantom Images and Modern Manifestations: Spirit Photography, Magic Theater, Trick Films, and Photography's Uncanny.' In *Fugitive images: From Photography to Video.* Edited by Patrice Petro, 42-71. Bloomington, IN: Indiana University Press, 1995.

Gunning, Tom. '"Primitive" Cinema: A Frame-up? Or the Trick's on Us.' *Cinema Journal* 28, no. 2 (1989): 3-12.

Gunning, Tom. 'Re-newing Old Technologies: Astonishment, Second Nature, and the Uncanny in Technology from the Previous Turn-Of-The-Century.' In *Rethinking Media Change: The Aesthetics of Transition.* Edited by Henry Jenkins and David Thornburn, 39-60. Cambridge, MA: MIT Press, 2003.

Gunning, Tom. 'The World as Object Lesson: Cinema Audiences, Visual Culture and the St. Louis World's Fair, 1904.' *Film History* 6, no. 4 (1994): 422-444.

Haber, Samuel. *Efficiency and Uplift: Scientific Management in the Progressive Era, 1890-1920.* Chicago, IL: University of Chicago Press, 1964.

Hall, Stuart, ed. *Representation: Cultural Representations and Signifying Practices.* New York: Sage, 1997.

Hatch, Mary Jo, and Michael Owen Jones. 'Photocopylore at Work: Aesthetics, Collective Creativity and the Social Construction of Organizations.' *Studies in Cultures, Organizations and Societies* 3, no. 2 (1997): 263-287.

Hampton, Mark. *Visions of the Press in Britain, 1850-1950.* Chicago, IL: University of Illinois Press, 2004.

Hansen, Mark B. N., and W. J. T. Mitchell, eds. *Critical Terms for Media Studies.* Chicago, IL: University of Chicago Press, 2010.

Hansen, Mark B. N. 'Media Theory.' *Theory, Culture & Society* 23, no. 2-3 (2006): 297-306.

Hauser, Ernst A. 'Raphael Eduard Liesegang 1869-1947' (Necrology). *Journal of Chemical Education* 26, no. 5 (May 1949): 274.

Haraway, Donna. *Primate Visions: Gender, Race, and Nature in the World Of Modern Science.* New York: Routledge, 1989.

Haraway, Donna. *Simians, Cyborgs and Women: The Reinvention of Nature.* New York: Routledge, 1991.

Hayles, N. Katherine. 'Virtual Bodies and Flickering Signifiers.' *October* 66 (1993): 69-91.

Hayles, N. Katherine. *How We Became Posthuman: Virtual Bodies in Cybernetics, Literature, and Informatics.* Chicago, IL: University of Chicago Press, 1999.

Hays, Samuel P. *Conservation and the Gospel of Efficiency: The Progressive Conservation Movement, 1890–1920.* Cambridge, MA: Harvard University Press, 1959.

Hempstead, Colin A. 'Representations of Transatlantic Telegraphy.' *Engineering Science & Education Journal* 4, no. 6 (1995): 17-25.

Hendricks, Gordon. *The Edison Motion Picture Myth.* Berkeley, CA: University of California Press, 1961.

Hendricks, Gordon. *The Kinetoscope: America's First Commercially Successful Motion Picture Exhibitor.* New York: Beginnings of the American Film, 1966.
Herbert, Stephen. *A History of Early Television*. New York: Taylor & Francis, 2004.
Herbert, Stephen. 'Professor Goaheadison's latest.' *Early Popular Visual Culture* 9, no. 1 (2011): 75-81.
Hidalgo, Santiago. 'Early American Film Publications: Film Consciousness, Self Consciousness.' In *A Companion to Early Cinema.* Edited by André Gaudreault, Nicolas Dulac, and Santiago Hidalgo, 202-219. New York: Wiley, 2012.
Hilmes, Michele. *Only Connect: A Cultural History of Broadcasting in the United States*. Boston, MA: Wadsworth Cengage Learning, 2013.
Hughes, Thomas. *Networks of Power: Electrification in Western Society, 1880-1930.* Baltimore, MD: Johns Hopkins University Press, 1993.
Hughes, Thomas. 'The Evolution of Large Technological Systems.' In *The Social Construction of Technological Systems: New Directions in the Sociology and History of Technology.* Edited by Trevor Pinch, Thomas Hughes, and Wiebe Bijker, 51-82. Cambridge, MA: MIT Press, 1987.
Huhtamo, Erkki, and Jussi Parikka, eds. *Media Archaeology: Approaches, Applications, and Implications*. Berkeley, CA: University of California Press, 2011.
Huhtamo, Erkki. 'Seeing at a Distance: Toward an Archaeology of the Small Screen.' In *Art@Science.* Edited by Christa Sommerer and Laurent Mignonneau, 262-278. New York: Springer, 1998.
Huhtamo, Erkki. 'Elements of Screenology: Toward an Archaeology of the Screen.' *ICONICS: International Studies of the Modern Image* 7 (2004): 31-82.
Huhtamo, Erkki. 'The Pleasures of the Peephole: An Archaeological Exploration of Peep Media.' In *The Book of Imaginary Media: Excavating the Dream of the Ultimate Communication Medium.* Edited by Eric Kluitenberg, 74-155. Rotterdam, The Netherlands: NAi Publishers, 2006.
Huhtamo, Erkki. 'Toward a History of Peep Practice.' *A Companion to Early Cinema.* Edited by André Gaudreault, Nicolas Dulac, and Santiago Hidalgo, 32-51. New York: Wiley, 2012.
Hunt, Verity. 'Electric Leisure: Late Nineteenth-Century Dreams of Remote Viewing by Telectroscope.' *Journal of Literature and Science* 7, no. 1 (2014): 55-76.
Ijsselsteijn, W. A. 'History of Telepresence.' In *3D Videocommunication: Algorithms, Concepts and Real-Time Systems in Human Centered Communication.* Edited by O. Schreer, P. Kauff, and T. Sikora, 7-21. New York: John Wiley & Sons, 2005.
Jacobson, Brian R. 'The Black Maria: Film Studio, Film Technology.' *History and Technology* 27, no. 2 (2011): 233-241.
Jacobson, Brian R. *Studios Before the System: Architecture, Technology, and Early Cinema.* New York: Columbia University Press, 2015.

Jenkins, Henry. 'What Happened Before Youtube?' In *YouTube: Online Video and Participatory Culture*. Edited by Jean Burgess and Joshua Green, 109-125. New York: John Wiley & Sons, 2013.

Jenkins, Reese Valmer, Charles Musser, and Thomas E. Jeffrey. *A Guide to motion picture catalogs by American producers and distributors, 1894-1908: a microfilm edition*. Frederick, MD: University Publications of America, 1985.

Johnston, Sean. 'The Construction of Colorimetry by Committee.' *Science in Context* 9 (1996): 387-420.

Johnston, Sean. *A History of Light and Colour Measurement: science in the shadows*. Bristol, UK: Institute of Physics Publishing, 2001.

Jordan, John M. *Machine-Age Ideology: Social Engineering and American Liberalism, 1911-1939*. Chapel Hill, NC: University of North Carolina Press, 1994.

Kane, Louise. '*To-Day* Has Never Been "Highbrow": Middlebrow, Modernism, and the Many Faces of To-Day.' In *Transitions in Middlebrow Writing, 1880-1930*. Edited by Kate Macdonald and Christopher Singer, 57-76. New York: Palgrave MacMillan, 2015.

Kaplan, E. Ann. *Looking for the Other: Feminism, Film, and the Imperial Gaze*. New York: Routledge, 1997.

Kern, Stephen. *The Culture of Time and Space, 1880-1918*. Cambridge, MA: Harvard University Press, 1983.

Kisseloff, Jeff. *The Box: An Oral History of Television, 1920-1961*. New York: Viking Press, 1995.

Kittler, Friedrich. *Gramaphone, Film, Typewriter*. Stanford, CA: Stanford University Press, 1999.

Kluitenberg, Eric, ed. *The Book of Imaginary Media: Excavating the Dream of the Ultimate Communication Medium*. Rotterdam, The Netherlands: NAi Publishers, 2007.

Kluitenberg, Eric. 'On the Archaeology of Imaginary Media.' In *Media Archaeology: Approaches, Applications, and Implications*. Edited by Erkki Huhtamo and Jussi Parikka, 48-69. Berkeley, CA: University of California Press, 2011.

Kroker, Arthur, and Marilouise Kroker, eds. *Critical Digital Studies: A Reader*. Toronto: University of Toronto Press, 2008.

Landon, Brooks. *The Aesthetics of Ambivalence: Rethinking Science Fiction Film in the Age of Electronic (Re)production*. Westport, CT: Greenwood Publishing Group, 1992.

Lant, Antonia. 'The Curse of the Pharaoh, or How Cinema Contracted Egyptomania.' *October* 59 (1992): 87-112.

Lastra, James. *Sound Technology and the American Cinema: Perception, Representation, Modernity*. New York: Columbia University Press, 2000.

Layard, George Somes. *The Life and Letters of Charles Samuel Keene.* London: Sampson Low, Marston and Co., 1892.

Leary, Patrick. *The Punch Brotherhood: Table Talk and Print Culture In Mid-Victorian London.* London: British Library, 2010.

Lee, Eun-Ju, and Soo Youn Oh. 'Computer-Mediated Communication.' In *Oxford Bibliographies in Communication*, 15 January 2015.

Levinson, Paul. *Digital McLuhan: A Guide to the Information Millennium.* New York: Routledge, 1998.

Lewis, A.V., and C. Nightingale, 'The Paradox of Videotelephony—Unconscious Assumptions and Undervalued Skills.' *BT technology journal* 17, no. 1 (1999): 47-58.

Light, Jennifer. 'Facsimile: A Forgotten "New Medium" from the 20th Century.' *New Media & Society* 8, no. 3 (2006): 355-378.

Lightman, Bernard. *Victorian Popularizers of Science: Designing Nature for New Audiences.* Chicago, IL: University of Chicago Press, 2009.

Lightman, Bernard. '"The Voices of Nature": Popularizing Victorian science.' In *Victorian Science in Context.* Edited by Bernard Lightman, 187-211. Chicago, IL: University of Chicago Press, 1997.

Lipartito, Kenneth. 'Picturephone and the Information Age: The Social Meaning of Failure.' *Technology and Culture* 44, no. 1 (2003): 50-81.

Loiperdinger, Martin. 'Lumiere's Arrival of the Train: Cinema's Founding Myth.' *The Moving Image* 4, no. 1 (2004): 89-118.

Lotz, Amanda. *The Television Will be Revolutionized.* New York: New York Universrsity Press, 2014.

Luckhurst, Roger. *Science Fiction.* Cambridge, UK: Polity, 2005.

Luthi, Max. *The European Folktale.* Bloomington, IN: Indiana University Press, 1982.

Magoun, Alexander. *Television: The Life Story of a Technology.* Westport, CT: Greenwood Publishing Group, 2007.

Maltby, Richard. 'New Cinema Histories.' In *Explorations in New Cinema History: Approaches and Case Studies.* Edited by Richard Maltby, Daniel Biltereyst, and Philippe Meers, 3-40. Malden, MA: Wiley-Blackwell, 2011.

Mair, Gordon. 'How Fiction Informed the Development of Telepresence and Teleoperation.' In *Virtual Augmented and Mixed Reality: Designing and Developing Augmented and Virtual Environments.* Edited by Randall Shumaker, 368-377. New York: Springer, 2013.

Marc, David, and Robert Thompson. *Television in the Antenna Age: A Concise History.* New York: John Wiley & Sons, 2008.

Marvin, Carolyn. 'The Electrical Imagination: Predicting the Future of Communications in Britain and the United States in the Late Nineteenth Century.' PhD diss., University of Illinois at Urbana-Champaign, 1979.

Marvin, Carolyn. *When Old Technologies Were New.* Cambridge, UK: Oxford University Press, 1988.

Marx, Leo. 'The Idea of "Technology" and Postmodern Pessimism.' In *Does Technology Drive History?: The Dilemma of Technological Determinism.* Edited by Merritt Roe Smith and Leo Marx, 237-258. Cambridge, MA: MIT Press, 1994.

Marx, Leo. *The Machine in the Garden: Technology and the Pastoral Ideal in America.* Cambridge, UK: Oxford University Press, 1964.

Massie, Keith, and Stephen D. Perry. 'Hugo Gernsback and Radio Magazines: An influential Intersection in Broadcast History.' *Journal of Radio Studies* 9 no. 2 (2002): 254-281.

Mayr, Otto. *The Origins of Feedback Control.* Cambridge, MA: MIT Press, 1970.

Mazlish, Bruce. *The Fourth Discontinuity: the Co-evolution of Humans and Machines.* New Haven, CT: Yale University Press, 1995.

McLean, Donald. *Restoring Baird's Image.* New York: Wiley-IEE, 2000.

McLuhan, Marshall. *Understanding Media: The Extensions of Man.* Cambridge, MA: MIT Press, 1964.

Mills, Mara. 'The Audiovisual Telephone: A Brief History.' In *Handheld? Music Video Aesthetics for Portable Devices.* Edited by Henry Keazor, 34-47. Heidelberg, Germany: ART-Dok, 2012.

Mills, Mara. 'Deafening: Noise and the Engineering of Communication in the Telephone System.' *Grey Room* 43 (2011): 118-143.

Mindell, David A. *Between Human and Machine: Feedback, Control, and Computing Before Cybernetics.* Baltimore, MD: Johns Hopkins University Press, 2002.

Mingers, John. *Realising Systems Thinking: Knowledge and Action in Management Science.* New York: Springer, 2006.

Mitcham, Carl. *Thinking Through Technology: The Path Between Engineering and Philosophy.* Chicago, IL: University of Chicago Press, 1994.

Mitchell, Timothy. 'The World as Exhibition.' *Comparative Studies in Society and History* 31, no. 2 (1989).

Mitchell, William J. *City of Bits: Space, Place, and the Infobahn.* Cambridge, MA: MIT Press, 1995.

Mitchell, W. J. T. *What Do Pictures Want?: The Lives and Loves of Images.* Chicago, IL: University of Chicago Press, 2005.

Mitchell, W. J. T. and Mark B. N. Hansen, eds. *Critical Terms for Media Studies.* Chicago, IL: University of Chicago Press, 2010.

Monteiro, Lois A. 'The Electronic Pocket Calculator: Joke 1.' *Western Folklore* 35, no. 1 (1976): 75.

Moore, Paul. 'Advance Newspaper Publicity for the Vitascope and the Mass Audience of Cinema's Reading Public.' In *A Companion to Early Cinema.* Edited by

Andre Gaudreault, Nicolas Dulac, and Santiago Hidalgo, 381-397. Hoboken, NJ: Wiley-Blackwell, 2012.

Moore, Paul. 'The Social Biograph: Newspapers as Archives of the Regional Mass Market for Movies.' In *Explorations in New Cinema History: Approaches and Case Studies.* Edited by Richard Maltby, Daniel Biltereyst, and Philippe Meers, 263-279. Hoboken, NJ: Wiley-Blackwell, 2011.

Morus, Iwan Rhys, ed. *Bodies/machines.* London: Bloomsbury Publishing, 2002.

Morus, Iwan Rhys. '"The Nervous System of Britain": Space, Time and The Electric Telegraph in the Victorian Age.' *The British Journal for the History of Science* 33, no. 4 (2000): 455-475.

Morus, Iwan Rhys. 'No Mere Dream: Material Culture and Electrical Imagination in Late Victorian Britain.' *Centaurus* 57, no. 3 (2015): 173-191.

Mumford, Lewis. *Technics and Civilization.* New York: Harcourt, 1934.

Mussel, James. *Science, Time and Space in the Late Nineteenth-Century Periodical Press.* Burlington, VT: Ashgate, 2007.

Musser, Charles. *Before the Nickelodeon: Edwin S. Porter and the Edison Manufacturing Company.* Berkeley, CA: University of California Press, 1991.

Musser, Charles. *Edison Motion Pictures, An Annotated Filmography 1890-1900.* Washington, D. C.: Smithsonian Institute Press, 1997.

Musser, Charles. *The Emergence of Cinema: The American Screen to 1907.* Berkeley, CA: University of California Press, 1994.

Musser, Charles. 'Movies and the Beginnings of Cinema,' in *American Cinema, 1890-1909: Themes and Variations.* Edited by Andre Gaudreault, 45-65. New Brunswick, NJ: Rutgers, 2009.

Natale, Simone, and Gabriele Balbi, 'Media and the Imaginary in History: The Role of the Fantastic in Different Stages of Media Change.' *Media History* 20, no. 2 (2014): 203-218.

Nead, Lynda. *Myths of Sexuality: Representations of Women in Victorian Britain.* Oxford: Blackwell, 1988.

Neubauer, John, ed. *Cultural History after Foucault.* Piscataway, NJ: Transaction Publishers, 1999.

Noakes, Richard. 'Punch and Comic Journalism in Mid-Victorian Britain.' In *Science in the Nineteenth-century Periodical: Reading the Magazine of Nature.* Edited by Geoffrey Cantor, 91-122. Cambridge, UK: Cambridge University Press, 2004.

Noakes, Richard. 'Representing "a Century of Inventions": Nineteenth-century Technology and Victorian *Punch.*' In *Culture and Science in the Nineteenth-Century Media.* Edited by L. Henson, G. Cantor, G. Dawson, et al., 151-163. Farnham, UK: Ashgate, 2004.

Noakes, Richard. 'Science in Mid-Victorian *Punch.*' *Endeavour* 26, no. 3 (2002): 92-96.

Noble, David F. *America by Design: Science, Technology, and the Rise of Corporate Capitalism.* Cambridge, UK: Oxford University Press, 1979.

Nye, David. *Electrifying America: Social Meanings of a New Technology, 1880-1940.* Cambridge, MA: MIT Press, 1992.

Nye, David. *American Technological Sublime.* Cambridge, MA: MIT Press, 1996.

Olsson, Jan. 'Framing Silent Calls: Coming to Cinematographic Terms with Telephony.' In *Allegories of Communication: Intermedial Concerns from Cinema to the Digital.* Edited by John Fullerton and Jan Olsson, 157-192. Bloomington, IN: Indiana University Press, 2004.

Orvell, Miles. *The Real Thing: Imitation and Authenticity in American Culture, 1880-1940.* Chapel Hill, NC: University of North Carolina Press, 1989.

Otis, Laura. 'The Other End of the Wire: Uncertainties of Organic and Telegraphic Communication.' *Configurations* 9, no. 2 (2001): 181-206;

Otter, Chris. *The Victorian Eye: A Political History of Light and Vision in Britain, 1800-1910.* Chicago, IL: University of Chicago Press, 2008.

Rank, Otto. 'The Myth of the Birth of the Hero.' In *In Quest of the Hero.* Edited by Robert Segal, 3-86. Princeton, NJ: Princeton University Press, 1990.

Parikka, Jussi. *What is Media Archaeology.* New York: Wiley, 2013.

Parikka, Jussi. 'Operative Media Archaeology: Wolfgang Ernst's Materialist Media Diagrammatics.' *Theory, Culture & Society* 28, no. 5 (2011): 52-74;

Perkowitz, Sydney. *Empire of Light: A History of Discovery in Science and Art.* New York: Henry Holt, 1996.

Peters, Benjamin. 'And Lead Us Not into Thinking the New is New: A Bibliographic Case for New Media History.' *New Media & Society* 11, no. 1-2 (2009): 13-30.

Pichler, Franz. 'Mechanisches Fernsehen: Fernsehempfang in der vorelektronischen Zeit' (Mechanical Television: Reception in the Pre-electronic Days). *Plus Lucis* 2 (2001): 21-26.

Poster, Mark. *What's the Matter with the Internet.* Minneapolis, MN: University of Minnesota Press, 2001.

Preston, Michael. 'Xerox-lore.' *Keystone Folklore* 19, no. 1 (1974): 11-26.

Preston, Michael. 'Traditional Humor from the Fax Machine: All of a Kind.' *Western Folklore* 53, no. 2 (1994): 147-169.

Prodger, Philip, and Tom Gunning. *Time Stands Still: Muybridge and the Instantaneous Photography Movement.* Cambridge, UK: Oxford University Press, 2003.

Propp, Vladimir. *Morphology of the Folktale.* Austin, TX: University of Texas Press, 1968.

Rabinbach, Anson. *The Human Motor: Energy, Fatigue, and the Origins of Modernity.* Berkeley, CA: University of California Press, 1992.

Rasmussen, Nicolas. *Picture Control: The Electron Microscope and the Transformation of Biology in America, 1940-1960.* Stanford, CA: Stanford University Press, 1999.

Reingold, Nathan. 'Definitions and Speculations: The Professionalization of Science in America in the Nineteenth Century.' In *The Pursuit of Knowledge in the Early American Republic.* Edited by Alexandra Oleson and Sanborn C. Brown. Baltimore, MD: Johns Hopkins University Press, 1976.

Rodowick, David. *The Virtual Life of Film.* Cambridge, MA: Harvard University Press, 2007.

Roggenkamp, Karen. *Narrating the News: New Journalism and Literary Genre in Late Nineteenth-Century American Newspapers and Fiction.* Kent, OH: Kent State University Press, 2005.

Rogers, Brett, and Benjamin Eldon Stevens, ed. *Classical Traditions in Science Fiction.* Cambridge, UK: Oxford University Press, 2015.

Rossell, Deac. *Living Pictures: The Origins of the Movies.* Albany, NY: State University of New York Press, 1998.

Rubinstein, Charles. 'Optics at Bell Laboratories – General Optics, television, and Vision.' *Applied Optics* 11, no. 11 (1972): 2401-2411.

Rydell, Robert. *World of fairs: the century-of-progress expositions.* Chicago, IL: University of Chicago Press, 1993.

Sandage, Scott. *Born Losers: A History of Failure in America.* Cambridge, MA: Harvard University Press, 2005.

Saunders, Barbara, and Jaap Van Brakel. 'The Trajectory of Color.' *Perspectives on Science* 10, no. 3 (2002): 302-355.

Schacker Jennifer. *National Dreams: The Remaking of Fairy Tales in Nineteenth-Century England.* Philadelphia, PA: University of Pennsylvania Press, 2004.

Schanda, Janos, ed. *Colorimetry: Understanding the CIE System.* New York: Wiley, 2007.

Sconce, Jeffrey. *Haunted Media: Electronic Presence from Telegraphy to Television.* Durham, NC: Duke University Press, 2000.

Secord, James. *Victorian Sensation: The Extraordinary Publication, Eeception, and Secret Authorship of 'Vestiges of the Natural History of Creation.'* Chicago, IL: University of Chicago Press, 2003.

Secord, James. *Visions of Science: Books and Readers at the Dawn of the Victorian Age.* Chicago, IL: University of Chicago Press, 2015.

Segal, Howard. *Technological Utopianism in American Culture.* Syracuse, NY: Syracuse University Press, 1985.

Segal, Howard. *Future Imperfect: The Mixed Blessings of Technology in America.* Amherst, MA: University of Massachusetts Press, 1994.

Shapin, Steven, and Simon Shaffer. *Leviathan and the Air-Pump: Hobbes, Boyle and the Experimental Life.* Princeton, NJ: Princeton University Press, 1985.

Shiers, George. *Early Television: A Bibliographic Guide.* London: Taylor & Francis, 1997.

Shiers, George. 'The Rise of Mechanical Television, 1901–1930.' *SMPTE Journal* 90, no. 6 (1981): 508-521.

Short, J. A., E. Williams, and B. Christie. *The Social Psychology of Telecommunications*. London: Wiley, 1976.

Snickars, Pelle, and Patrick Vonderau, eds. *The YouTube Reader*. Stockholm: National Library of Sweden, 2009.

Solomon, Matthew. *Fantastic Voyages of the Cinematic Imagination: Georges Méliès's Trip to the Moon*. Albany, NY: State University of New York Press, 2011.

Sorin-George, Toma, Ana-Maria Grigore, and Paul Marinescu. 'The Emergence of Scientific Management in America.' *Manager* 19 (2014): 128-131.

Spehr, Paul. 'Filmmaking at the American Mutoscope and Biograph Company 1900—1906.' *The Quarterly Journal of the Library of Congress* 37, no. 3-4 (Summer/Fall 1980): 413-421.

Spehr, Paul. *The Man Who Made Movies: W.K.L. Dickson*. New Barnet, UK: John Libbey Publishing, 2008.

Spielmann, Marion. *The History of Punch*. London: Cassell Publishing, 1895.

Spigel, Lynn. *Make Room for TV: Television and the Family Ideal in Postwar America*. Chicago, IL: University of Chicago Press, 1992.

Stafford, Barbara Maria, Frances Terpak, and Isotta Poggi. *Devices of Wonder: From the World in a Box to Images on a Screen*. Los Angeles, CA: Getty Publications, 2001.

Staiger, Janet. *Media Reception Studies*. New York: New York University Press, 2005.

Standage, Tom. *The Victorian Internet: The Remarkable Story of the Telegraph and the Nineteenth Century's Online Pioneers*. London: Weidenfeld & Nicolson, 1998.

Sterling, Christopher H., and John Michael Kittross. *Stay Tuned: A History of American Broadcasting*. New York: Routledge, 2001

Sterne, Jonathan, and Dylan Mulvin. 'The Low Acuity for Blue: Perceptual Technics and American Color Television.' *Journal of Visual Culture* 13, no. 2 (2014): 118-138.

Sterne, Jonathan. *The Audible Past: Cultural Origins of Sound Reproduction*. Durham, NC: Duke University Press, 2003.

Sterne, Jonathan. *MP3: The Meaning of a Format*. Durham, NC: Duke University Press, 2012.

Stiegler, Bernard. *Technics and Time: The Fault of Epimetheus*. Stanford, CA: Stanford University Press, 1998.

Stollen, Hugh. *Radio and Television Regulation: Broadcast Technology in the United States*. Baltimore, MD: Johns Hopkins University Press, 2000.

Strauven, Wanda, ed. *The Cinema of Attractions Reloaded*. Amsterdam: Amsterdam University Press, 2006.

Sturken, Marita, Douglas Thomas, and Sandra Ball-Rokeach. *Technological Visions: The Hopes and Fears that Shape New Technologies*. Philadelphia, PA: Temple University Press, 2004.

Sumner, David E. *The Magazine Century: American Magazines Since 1900*. Bern: Peter Lang, 2010.

Sussman, Herbert. *Victorian Technology: Invention, Innovation, and the Rise of the Machine*. Santa Barbara, CA: Praeger, 2009.

Telotte, J. P. *A Distant Technology: Science Fiction Film and the Machine Age*. Middletown, CT: Wesleyan University Press, 1999.

Telotte, J.P. 'Film and/as Technology.' *Post Script* 10 (1990): 3-8.

Thomas, Julia. *Pictorial Victorians: The Inscription of Values in Word and Image*. Athens, OH: Ohio University Press, 2004.

Thompson, Kenneth. 'Introduction to the Early Sociology of Management and Organizations.' In *Scientific Management*. New York: Routledge, 2003.

Thorburn, David, and Henry Jenkins, eds. *Rethinking Media Change: The Aesthetics of Transition*. Cambrige, MA: MIT Press, 2003.

Toscano, Aaron. *Marconi's Wireless and the Rhetoric of a New Technology*. New York: Springer, 2012.

Tosi, Virgilio. *Cinema Before Cinema: The Origins of Scientific Cinematography*. Translated by Sergio Angelini. London: British Universities Film and Video Council, 2005.

Turkle, Sherry. *Alone Together: Why We Expect More from Technology and Less from Each Other*. New York: Basic Books, 2011.

Uricchio, William, and Roberta Pearson. 'Coming to Terms with New York City's Moving Picture Operators, 1906–1913.' *The Moving Image: The Journal of the Association of Moving Image Archivists* 2, no. 2 (2002): 73-93.

Uricchio, William. 'Historicizing Media in Transition.' In *Rethinking Media Change: The Aesthetics of Transition*. Edited by David Thorburn and Henry Jenkins, 23-38. Cambridge, MA: MIT Press, 2003.

Uricchio, William. 'Phantasia and Technè at the Fin-de-siècle.' *Intermediality: History and Theory of the Arts, Literature and Technologies* 6 (2005): 27-42.

Uricchio, William. 'Storage, Simultaneity, and the Media Technologies of Modernity.' In *Allegories of Communication: Intermedial Concerns from Cinema to the Digital*. Edited by John Fullerton and Jan Olsson, 123-138. Bloomington, IN: Indiana University Press, 2004.

Uricchio, William. 'Television, Film and the Struggle for Media Identity.' *Film History* 10, no. 2 (1998): 118-127.

Uricchio, William. 'Television's First Seventy-Five Years: The Interpretive Flexibility of a Medium in Transition.' In *The Oxford Handbook of Film and Media Studies*. Edited by Robert Kolker, 286-305. Cambridge, UK: Oxford University Press, 2008.

Uricchio, William. 'The Future of a Medium Once Known as Television.' In *The YouTube Reader*. Edited by Patrick Vonderau and Pelle Snickars, 24-39. Stockholm: National Library of Sweden, 2009.

Väliaho, Pasi. *Mapping the Moving Image: Gesture, Thought and Cinema Circa 1900*. Amsterdam: Amsterdam University Press, 2010.
Van den Ende, Jan, Wim Ravesteijn, and Dirk De Wit. 'Shaping the Early Development of Television.' *Technology and Society Magazine, IEEE* 16, no. 4 (1997): 13-26.
Van der Vleuten, Erik. 'Large Technical Systems.' In *A Companion to the Philosophy of Technology*. Edited by Jan Olsen, Stig Andur Pedersen, and Vincent F. Hendricks, 218-222. Malden, MA: John Wiley & Sons, 2012.
Van Heur, Bas, Loet Leydesdorff, and Sally Wyatt. 'Turning to Ontology in STS? Turning to STS Through "Ontology".' *Social Studies of Science* 43, no. 3 (2012): 341-362.
Van Loon, Joost. *Media Technology: Critical Perspectives*. New York: McGraw Hill, 2008.
Varnava, Andrekos. 'Punch and the British Occupation of Cyprus in 1878.' *Byzantine and Modern Greek Studies* 29, no. 2 (2005): 167-186.
Virilio, Paul. *The Vision Machine*. Bloomington, IN: Indiana University Press, 1994.
Voskuhl, Heidi. 'Engineering Philosophy: Theories of Technology, German Idealism, and Social Order in High-Industrial Germany.' *Technology and Culture* 57, no. 3 (2016).
Wade, Nicholas. *Visual Perception: An Introduction*. New York: Psychology Press, 2013.
Weber, Anne-Katrin. 'Recording on Film, Transmitting by Signals: The Intermediate Film System and Television's Hybridity in the Interwar Period.' *Grey Room* 56 (2014): 6-33.
Weis, Elisabeth, and John Belton. *Film Sound: Theory and Practice*. New York: Columbia University Press, 1985.
Willems, Philippe. 'A Stereoscopic Vision of the Future: Albert Robida's Twentieth Century.' *Science Fiction Studies* 26, no. 3 (1999): 354-378.
Williams, Raymond. *Television: Technology and Cultural Form*. New York: Routledge, 1974.
Willis, Martin. *Vision, Science and Literature, 1870-1920: Ocular Horizons*. New York: Routledge, 2015.
Wilson, Catherine. *The Invisible World: Early Modern Philosophy and the Invention of the Microscope*. Princeton, NJ: Princeton University Press, 1995.
Winkler, Stefan. *Digital Video Quality: Vision Models and Metrics*. New York: John Wiley & Sons, 2005.
Winner, Langdon. 'Upon Opening the Black Box and Finding It Empty: Social Constructivism and the Philosophy of Technology.' *Science, Technology, & Human Values* 18, no. 3 (1993): 362-378.
Winston, Brian. *Misunderstanding Media*. New York: Routledge, 1986.
Winston, Brian. *Media, Technology and Society*. New York: Routledge, 1998.

Woolgar, Steve, and Javier Lezaun. 'The Wrong Bin Bag: A Turn to Ontology in Science and Technology Studies?' *Social Studies of Science* 43, no. 3 (2013): 321-340.

Wu, Hong Ren, and Kamisetty Ramamohan Rao, eds. *Digital Video Image Quality and Perceptual Coding.* Boca Raton, FL: CRC Press, 2005.

Wurtzler, Steve. *Electric Sounds: Technological Change and the Rise of Corporate Mass Media.* New York: Columbia University Press, 2007.

Wythoff, Grant. 'Pocket Wireless and the Shape of Media to Come, 1899–1922.' *Grey Room* 51 (2013): 40-63.

Young, Paul. 'Media on Display: A Telegraphic History of Early American Cinema.' In *New Media, 1740-1915.* Edited by Lisa Gitelman and Geoffrey B. Pingree, 229-264. Cambridge, MA: MIT Press, 2004.

Yuste, Antonio Perez, 'La televisión mecánica' (The Mechanical Television). In *Detrás de la Cámara. Historia de la Televisión y de sus Cincuenta Años en España (Behind the Scenes: History of Television and its Fifty Years in Spain)*, 65-82. Madrid: Colegio Oficial de Ingenieros de Telecomunicación, 2008.

Zajonc, Arthur. *Catching the Light: The Entwined History of Light and Mind.* Cambridge, UK: Oxford University Press, 1995.

Zeise, Tina. *Geschichte und Technik des analogen Fernsehens (History and technology of Analog Television).* Munich: GRIN Verlag, 2006.

Zielinski, Siegfried. *Audiovisions: Cinema and Television as Entr'actes in History.* Amsterdam: Amsterdam University Press, 1999.

Zylinska, Joanna, ed. *The Cyborg Experiments: The Extensions of the Body in the Media Age.* London: A&C Black, 2002.

Primary Sources [Chronological]

19th Century History of Technology and Culture

Archer, Charles Maybury, ed. *The London Anecdotes for All Readers.* London: D. Bogue, 1848.

New Monthly Magazine. 'Romance of the Electric Telegraph.' June 1850, 296-307.

Chambers' Edinburgh Journal. 'An Evening with the Telegraph.' January 1851, 1-4.

Preston, George Bartlett. *History, Theory and Practice of the Electric Telegraph.* Boston: Ticknor and Fields, 1866.

Lardner, Dionysius. *The Electric Telegraph.* London: James Walton, 1867.

Timbs, John. *Wonderful Inventions: From the Mariner's Compass to the Electric Telegraph Cable.* London: George Routledge and Sons, 1868.

Sabine, Robert. *The History and Progress of the Electric Telegraph.* New York: D. Van Nostrand, 1869.

Baile, Jean Baptiste Alexander. *Wonders of Electricity.* New York: Scribner, Armstrong and Co., 1872.

Harpers New Monthly Magazine. 'The Telegraph.' August 1873, 332-360.

Deschanel, A. *Electricity and Magnetism.* New York: Appleton, 1876.

Routledge, Robert, and John Henry Pepper. *Discoveries and Inventions of the Nineteenth Century.* London: George Routledge and Sons, 1876.

Kapp, Ernst. *Grundlinien einer Philosophie der Technik.* Braunschweig, Germany: Druck und Verlag von George Westermann, 1877.

Johnston, W. J., ed. *Lightning Flashes and Electric Dashes: A Volume of Choice Telegraphic Literature, Humor, Fun, Wit & Wisdom.* New York: W. J. Johnston, 1877.

Babcock, W. H. 'The Future of Invention.' *The Atlantic Monthly,* August 1879, 137-146.

Johnston, W. J., ed. *Telegraphic Tales and Telegraphic History.* New York: W. J. Johnston, 1880.

Thayer, Ella Cheever. *Wired Love: A Romance of Dots and Dashes.* New York: W. J. Johnston, 1880.

Hammond, Robert. *The Electric Light in our Homes.* London: Frederick Warne and Co., 1884.

Hale, Edward Everett. *Stories of Invention, Told by Inventors and Their Friends.* Boston, MA: Roberts Brothers,, 1885.

Burnley, James. *The Romance of Invention: Vignettes from the Annals of Industry and Science.* London: Cassell and Co, 1886.

Moskowitz, Sam, ed. *The Crystal Man: Stories by Edward Page Mitchell (1874-1886).* New York: Doubleday, 1974.

Mendenhall, Thomas C. *A Century of Electricity.* London: Macmillan, 1887.

Bellamy, Edward. *Looking Backward: 2000-1887.* New York: Ticknor & Co., 1888.

Bellamy, Edward. 'How I Came to Write Looking Backward.' *The Nationalist* (Boston). May 1889, 1-4.

Harris, William. 'Edward Bellamy's Vision.' *The Forum.* September 1889.

Verne, Jules, and Michel Verne. 'In the Year 2889.' *Forum* September 1888-February 1889.

McClure, J. B. *Edison and his Inventions.* Chicago, IL: Rhodes and McClure Publishing Co., 1889.

Stieringer, Luther. *The Life and Inventions of Thos. A. Edison.* New York: C. G. Burgoyne, 1890.

Jerrod, Walter. *Electricians and their Marvels.* London: S. W. Partridge 1890.

Caillard, Emma Marie. *Electricity, The Science of the Nineteenth Century, A Sketch for General Readers.* New York: D. Appleton and Co., 1891.

Liesegang, R. E. *Beiträge zum Problem des elektrichen Fernsehen.* Dusseldorf, 1891; 1899.

Ramey, Sanford. *Triumphs of Genius.* Philadelphia, PA: A. R. Keller Co., 1893.

Freeman's Journal and Commercial Advertiser. 'The Future of Invention.' 16 June 1893.

Ramey, Sanford. *Triumphs of Genius.* Philadelphia, PA: A. R. Keiler, 1893.

Flammarion, Camille. *La Fin Du Monde* (*Omega: The Last Days of the World*). Paris: Flammarion, 1894.

Cochrane, Robert. *The Romance of Industry and Invention.* Philadelphia, PA: J. B. Lippincott, 1896.

Cochrane, Charles Henry. *The Wonders of Modern Mechanism.* Philadelphia, PA: J. B. Lippincott, 1896.

Mee, Arthur. 'The Pleasure Telephone.' *The Strand Magazine,* September 1898, 339-345.

Baker, R.S. *The Boy's Book of Inventions; Stories of the Wonders of Modern Science.* New York: Doubleday & McClure, 1899.

Boyd, James. *Triumphs and Wonders of yhe 19th Century, The True Mirror of a Phenomenal Era.* Philadelphia, PA: A. J. Holman & Co: 1899.

Bidwell, Shelford. *Curiosities of Light and Sight.* London: Swann, Sonnenschein and Co., Ltd., 1899.

Iles, George. *Flame, Electricity and the Camera: Man's Progress from the First Kindling of Fire to the Wireless Telegraph and the Photography Of Color.* New York: Doubleday & McClure, 1900.

Byrn, Edward. *The Progress of Invention in the Nineteenth Century.* London: Munn and Co., 1900.

Steele, Penn. 'Anticipations: How Some of Mr. Wells' Speculative Predecessors have Fared.' *The Era Magazine,* January-June 1902, 460-468.

Cochrane, Charles Henry. *Modern Industrial Progress.* Philadelphia, PA: J. B. Lippincott, 1904.

Edisoniana

New York Times. 'Edison's Wonderful Inventions.' 15 March 1878.

New York World. 'Inventor Edison's Last.' 21 March 1878.

New York Sun. 'Talking of Futurity.' 24 March 1878.

New York Times. 'The Aerophone.' 25 March 1878.

Washington Herald (reprinted from *San Francisco Post*). 'Miseries of the Phonograph.' 1 April 1878.

Edison, Thomas Alva. Technical Notes and Drawings, 2 April 1878 (NV17065; TAEM 4:931).

Puck. 'That Awful Phonograph.' 24 April 1878.

New York Times. 'Practical Uses of the Phonograph.' 28 April 1878.

New York Herald. 'Edison's Wonder.' 28 April 1878.

New York Herald. 'The Future of the Phonograph.' 28 April 1878.

Daily Graphic. 'Invasions by the Telephone.' 30 April 1878.
Preston, George Bartlett. 'The Telephone and the Phonograph.' *Scribner's*, April 1878, 848-858.
Edison, Thomas. 'The Phonograph and Its Future.' *North American Review*, May 1878, 527-537.
Daily Graphic. 'Edisoniana.' 9 May 1878.
New York Tribune. 'Untitled' (This discovery mania). 14 May 1878.
Edison, Thomas Alva, and James Edward Kelly. Technical Notes and Drawings. April-May 1878: 4-6 (B037BA; TAEM 0:0).
Baltimore American. 'The Edison Phonograph.' 21 May 1878.
The Daily Graphic. 'Ears for the Deaf.' 5 June 1878.
The Daily Graphic. 'The Latest Offerings of Science.' 8 June 1878.
New York Sun. 'Edison's "Ear Telescope".' 8 June 1878.
New York Sun. 'Edison Should Make Haste to Abandon the Terrific Name' (The Telescopophone). 11 June 1878.
The Nation. 'Notes' (The Telescopophone). 13 June 1878, 387.
Lowell Daily Citizen (Lowell, MA). 'In his Search for an Ear Trumpet.' 15 June 1878.
Indianapolis Herald. 'The Amusement Record.' 15 June 1878.
Milwaukee Daily Sentinel (Milwaukee, WI). 'Another Edisonism.' 17 June 1878.
Christian Union. 'Fact and Rumor.' 19 June 1878.
Daily Evening Bulletin (San Francisco, CA). 'Edison's Megaphone.' 20 June 1878.
Times (London). 'The Megaphone—A Remedy for Deafness.' 27 June 1878.
Illustrated Science News. 'Edison's Marvelous Inventions.' 1 July 1878, 103.
English Mechanic. 'The Phonoscope.' 5 July 1878, 411.
New York Sun. 'Edison Outdoing Himself.' 11 July 1878.
The Daily Graphic. 'Edison's Latest Inventions; More About Edison's Wonders.' 19 July 1878.
Fun. 'The Long and Short of It.' 19 June 1878.
Examiner (UK). 'The Latest Scientific Squabble.' 27 July 1878.
New York Tribune. 'The Megaphone.' 17 August 1878.
Scientific American. 'Edison's Megaphone.' 24 August 1878, 111, 114.
Engineering and Mining Journal. 'Edison, The Inventor of the Phonograph.' August 1878, 131.
New York Sun. 'Invention's Big Triumph.' 10 September 1878.
'General Notes: The Megaphone.' *Journal of the Society of Arts* (September 1878): 902.
Electric Review. 'Edison's Megaphone' (reprinted from *Scientific American*). 15 September 1878, 383-384.
New York Sun. 'Edison's Newest Marvel.' 16 September 1878.
New York Sun. 'Power Flashed by Wire.' 17 September 1878.
Blount, John, letter to the editor, *Times* (London), 4 October 1878.

Hollingshead, John, letter to the editor, *Times* (London), 5 October 1878.
Clarke, Somers, letter to the editor, *Times* (London), 7 October 1878.
Terrault, S. E., letter to the editor, *Times* (London), 8 October 1878.
M. O., letter to the editor, *Times* (London), 8 October 1878.
L. W., letter to the editor, *Times* (London), 9 October 1878.
The Standard (London). 'The New Electric Light.' 9 October 1878.
The Daily News (London). 'The Electric Light.' 9 October 1878.
Reed, E. J. letter to the editor, *Times* (London), 10 October 1878.
Liang, R. J., letter to the editor, *Times* (London), 10 October 1878.
G. W., letter to the editor, *Times* (London), 10 October 1878
W. M. A., letter to the editor, *Times* (London), 11 October 1878.
Manchester Weekly Times. 'The Electric Light for Domestic Purposes.' 12 October 1878.
New York Herald. 'Edison's Electric Light.' 12 October 1878.
Squire, W. S., letter to the editor, *Times* (London), 12 October 1878.
Weekly Dispatch. 'The Electric Light.' 13 October 1878.
Tyler, H. W., letter to the editor, *Times* (London), 14 October 1878.
Times (London), 'The Correspondence we have recently published.' 14 October 1878.
The Standard (London). 'The Electric Light.' 15 October 1878.
Saturday Review of Politics, Literature, Science and Art. 'American Literature.' 16 October 1878, 539-541.
Examiner. 'The Future of Gas.' 16 October 1878.
New York Sun. 'Electric and Gas Light.' 18 October 1878.
Adams, James, letter to the editor, *Times* (London), 18 October 1878
Siemens, C. W. letter to the editor, *Times* (London), 19 October 1878.
Squire, W. S. letter to the editor, *Times* (London), 19 October 1878.
Saturday Review of Politics, Literature, Science and Art.'Electricity and Gas.' 19 October 1878, 481-482.
New York Sun. 'Light from Electricity.' 19 October 1878.
Manchester City News. 'The Electric Light and Mr. Edison's Discovery.' 19 October 1878.
Punch. 'Ode to the Coming Light.' 19 October 1878.
New York Sun. 'Edison's Electric Light.' 20 October 1878.
Daily Graphic. 'Edison's New Light.' 21 October 1878.
Daily Graphic. 'The New Inventions.' 21 October 1878.
Times (London) (from *New York Herald*). 'Edison's Electric Light.' 22 October 1878.
New York Times. 'London Talk and Gossip: Matters that Occupy English Minds.' 25 October 1878.
Daily Graphic. 'The Man Who Moves the World.' 26 October 1878.
Punch. 'The Edison Light, and the Silly Birds.' 26 October 1878.

The Electrician. 'Edison's Electric Light.' 26 October 1878.

Times (London). 'The Electric Light.' 26 October 1878.

Examiner. 'The Future of Gas.' 26 October 1878.

New York Times. 'Gas Stocks and Light.' 27 October 1878.

New York Times. 'The Electric Light.' 30 October 1878.

New York Sun. 'Light for London: Gas and the Electric Light.' 30 October 1878.

'Edison's Inventions.' *Nature* 18 no. 469 (October 1878): 674-676.

Gourard, George, letter to the editor, *Times* (London), 1 November 1878.

Cassels, James Patterson, 'The Megaphone,' letter to the editor, *British Medical Journal,* 2 November 1878, 679.

New York Times. 'Scientific Gossip.' 3 November 1878.

New York Times. 'Uneasiness in England: What Occupies the London Mind.' 3 November 1878.

New York Times. 'What London Talks About.' 5 November 1878.

Times (London). 'The Megaphone.' 11 November 1878.

Fun. 'Electric Light Memo's.' 13 November 1878.

Fun. 'Re the Glasgow Bank.' 13 November 1878.

The Metropolitan (London). 'The Electric Light.' 16 November 1878.

Fun, 'The Scientific Age.' 20 November 1878.

Punch. 'Brightness and Beauty.' 23 November 1878.

Bishop, William H. 'A Night with Edison.' Illustrations by James E. Kelly. *Scribner's monthly,* November 1878, 88-99.

Times (London). 'The Electric Light.' 6 December 1878.

Punch. 'The Sorceries of Science.' 7 December 1878.

New York Times. 'Progress of the Electric Light.' 8 December 1878.

Punch. 'In the Light of the Future or, How We Shall Have to Talk.' 9 December 1878.

Punch. 'Recent Improvements in Science' (The Chrysophone).' 14 December 1878.

Punch. 'Edison's Anti-Gravity Underclothing.' 9 December 1878.

Punch. 'Edison's Telephonoscope.' 9 December 1878.

Expectant, letter to the editor, *Times* (London), 20 December 1878.

Judy, or the London Seriocomic Journal. 'The Coming Light (Electrophote).' 24 December 1878.

New York Times. 'Edison's Electric Light: Conflicting Statements as to its Utility.' 28 December 1878.

Times (London). 'Mr. Edison.' 31 December 1878.

Donisthorpe, W. 'Talking Photographs,' letter to the editor, *Nature* 17, no. 430 (1878): 242.

Garbit, Frederick. *The Phonograph and Its Inventor, Thomas Alvah Edison.* Boston: Gunn, Bliss and Co., 1878.

New York Times. 'Edison and the Skeptics.' 4 January 1879.

Fun. 'Our Extra-Special and Mr. Edison.' 8 January 1879.

Punch. 'Edisoniana.' 18 January 1879.

Lancaster, W. J. 'Edison's Megaphone,' letter to the editor, *English Mechanic,* 24 January 1879, 501.

Time. 'A Visit to Professor Edison.' April 1879.

New York Times. 'Edison and the Skeptics.' 4 January 1880.

Puck. 'Edison!' 4 February 1880.

The Photographic Times. 'Notes and News' (Something New). 22 June 1888, 296.

New York World. 'Edison's Talking Baby.' 23 June 1888.

Times (Pittsburg). 'Untitled' (Edison uninterested in 'increasing the human vision'). 30 July 1888.

Edison, Thomas. Motion Picture Patents and Caveats. 8 October 1888. Signed 15 October 1888, filed 17 October 1888. (PT031AAA; TAEM 113:236).

Edison, Thomas. Motion Picture Patents and Caveats. 25 March 1889. (QM001354; TAEM 116:196).

Electrical Review. 'How Electricity Will Help the Editor of the Future.' 2 February 1889, 4.

Townsend, Horace. 'Edison: His Work and his Workshop.' *Cosmopolitan* (April 1889), 598-607: 602.

Edison, Thomas. Telegraph, Telephone and Phonograph patents and caveats. 12 May 1889. (PT031AAF1; TAEM 113:497).

Boston Journal. 'Edison's Last.' 13 May 1889.

The Republican (Meridan, CT). 'Untitled.' (Edison's Far-Sight Machine). 14 May 1889.

Far-Sight Machine, Menlo Park Scrapbook, 1889 (TAED SM035).

The American (Waterbury, CT). 'Untitled' (Edison's Far-Sight Machine). 22 May 1889.

Milwaukee Daily Journal (Milwaukee, WI). 'Personal and Impersonal' (If Edison had his Far-sight machine in order). 24 May 1889.

The World (Omaha, NE). 'Untitled' (Edison's Far-Sight Machine). 24 May 1889.

Electrical Review. 'A Far-Sight Machine.' 25 May 1889.

Lincoln Journal (Lincoln, NE). 'Untitled' (Edison's Far-Sight Machine). 26 May 1889.

Scientific American. 'A Far-Sight Machine.' 1 June 1889.

Baltimore Herald. 'Untitled (Edison's Far-Sight Machine).' 10 June 1889.

The Statesman (Yonkers, NY). 'A Far-Sight Machine.' 11 June 1889.

St. James Gazette. 'Was Mr. Edison in Earnest?' 12 June 1889.

Puck. 'It Comes Too Late.' 12 June 1889.

Iron. 'What Next?' 14 June 1889.

Aberdeen Weekly Journal. 'Edison's Coming Invention.' 15 June 1889.

Western Mail (UK). 'Edison's Far-Sight Machine.' 15 June 1889.

Birmingham Daily Post. 'Gleanings' (One of the most peculiar of Mr. Edison's recent inventions). 15 June 1889.

Pall Mall Gazette. 'Fourth Edition' (Edison's Latest—A Far-Sight Machine). 15 June 1889.
Lloyd's Weekly Newspaper (UK). 'Edison's Latest Invention.' 16 June 1889.
The Post (Liverpool, England). 'A Far-Sight Machine.' 17 June 1889.
The Record (West Chester, PA). 'Untitled' (Edison's Far-Sight Machine). 18 July 1889.
Commercial Gazette (Cincinatti, OH). 'Untitled' (Edison's Far-Sight Machine). 18 July 1889.
The Journal (Pottsville, PA). 'Untitled' (Edison's Far-Sight Machine). 19 June 1889.
Dalin to TAE, 22 June 1889 (TAED D8923AAQ).
The Courier and London & Middlesex Counties Gazette. 'A Far-Sight Machine.' 22 June 1889.
Yenowine's News (Milwaukee, WI). 'Gossip on the Quiet.' 23 June 1889.
Tribine (Racine, OH). 'Untitled' (Edison's Far-Sight Machine). 24 July 1889.
New York Times. 'A World's Fair in 1892.' 25 June 1889.
The Time (Bethlehem, PA). 'Untitled' (Edison's Far-Sight Machine). 25 June 1889.
Judy. 'A Run on 'Far-Sight Machines.' 26 June 1889.
Judy. 'Pepper and Salt.' 26 June 1889.
Judy. 'The Twigger.' 26 June 1889.
New York Graphic. 'Untitled' (Edison's Far-Sight Machine.' 26 June 1889.
Fraser to TAE, 11 July 1889 (TAED D8923AAR).
Fun. 'Professor Goaheadison's Latest.' 3 July 1889.
Fun. 'Goaheadison's Real Latest.' 17 July 1889.
The Times (Portsmouth, NH). 'Untitled' (Edison's Far-Sight Machine). 29 July 1889;
The Electrical Engineer. 'Electro-photography.' July 1889, 329-330.
Electrical Review. 'A Far-sighted Individual.' 2 August 1889, 126.
The Leader (Wilkes Barre, PA). 'Untitled' (Edison's Far-Sight Machine). 9 August 1889.
Atchison Daily Champion (Atchison, KS). 'Edison's Machines.' 10 August 1889.
Punch. 'Open House (To be dated after the next invention).' 10 August 1889.
Telegraphic Journal and Electrical Review. 'The Telephote' (Courtonne). 16 August 1889.
The Democrat (Hutchinson, KS). 'Untitled' (Edison's Far-Sight Machine). 17 August 1889.
The Illustrated London News (American edition, NY). 'Untitled.' 19 August 1889.
Washington Post. 'The American Wizard: More Wonderful Things that Edison is to Bring Forth.' 1 September 1889.
Levant Herald. 'Mr. Edison and the Electric Millennium.' 1 September 1889.
The Commercial (Portland, IN). 'Untitled' (Edison's Far-Sight Machine). 5 September 1889.
Detroit Free Press. 'Remnants – Some Wonderful Inventions.' 6 October 1889.
New York Herald. 'Mr Edison at Home Unspoiled by Glory.' 7 October 1889.

Brooklyn Journal. 'Edison's Talk.' 7 October 1889.

Lockwood, Thomas D. 'Observations.' *The Electrical Engineer* (US). October 1889, 423.

Edison, Thomas. 'The Dangers of Electric Lighting.' *The North American Review.* November 1889, 625-635.

Harper's Weekly. 'The Electrical Wire Peril.' 14 December 1889.

Iron. 'The Past Year. A General Retrospective.' 3 January 1890.

Electrical World. 'Special Correspondence: Personal' (The Electrical Barnum), 11 January 1890, 97.

Sacramento Daily Record. 'Latest from the Wizard.' 1 February 1890.

New York Herald. 'To Catch a Speaker's Gestures.' 2 February 1890.

Sacramento Daily Record. 'Edison's Latest.' 3 February 1890.

Pall Mall Gazette (London). 'Untitled (Edison has added a new horror to existence).' 3 February 1890.

Morning Advertiser (London). 'Untitled' (With all its 'hem's and 'ha's). February 4, 1890.

Puck. 'What Next?' 5 February 1890.

New York Star. 'Popular Dolls; A Rush for Edison's Phonographic Toys' (reprinted in *The Washington Critic*). 6 February 1890.

Lathrop, George Parsons. 'Finding Nature's Secrets: A Day in the Life of Edison, the Wizard of Menlo Park.' *New York World.* 7 February 7, 1890.

Wahpeton Times (Wahpeton, ND). 'A Dead Madman's Words.' 13 February 1890.

News and Citizen (Morrisville, VT). 'Untitled' (One of the curiosities of Edison). 13 February 1890.

The Atlanta Constitution (Atlanta, GA). 'People Here and There.' 14 February 1890.

Perrysberg Journal. 'Untitled' (Edison at the Electric Light Convention). 15 February 1890.

Lathrop, George Parsons. 'Talks with Edison.' *Harper's.* February 1890, 425-435.

Barnum, P. T. 'Notes and Comments: What the Fair Should Be.' *North American Review.* March 1890, 400-401.

Milwaukee Daily Journal (Milwaukee, WI). 'Wizard of Menlo Park.' 26 April 1890.

The Big Sandy News (Louisa, KY). 'The Kinetograph: With it you can see a man a thousand miles away—Edison's Latest.' 28 April 1890.

American Engineer. 'Engineering and Manufacturing Notes.' 7 May 1890.

Electrical Engineer. 'Notes: Electric Seeing.' 30 May 1890.

American Settler. 'Chat.' 12 July 1890.

The Electrical World. 'A Remarkably Successful Joke' (Preece-Hughes hoax). 9 August 1890, 82.

New York Sun. 'To See by Electricity: An Apparatus for the Eye as the Phonograph is for the Ear' (attributed to *Iron*). 12 September 1890.

Illustrated American. 'Between the Acts.' 4 October 1890, 3.

Morning Oregonian. 'Electric Marvels: An Apparatus for the eye as the phonograph is to the ear.' 16 November 1890.

The Chicago Evening Post. 'Edison's in Chicago: The Wizard of Menlo Park Stopping at the Auditorium, Tells of his Latest Invention.' 12 May 1891.

Chicago Journal. 'What Edison Claims.' 13 May 1891.

New York Sun. 'Edison's Conjury.' 13 May 1891.

Chicago Tribune. 'Edison and the Big Fair.' 14 May 1891.

The Times (London). 'Mr. Edison and the Chicago Exhibit.' 14 May 1891.

Electrical Engineer. 'Edison's Photophonokinetograph,' 20 May 1891, 584.

English Mechanic. 'Edison's Photophonokinetograph.' 20 May 1891.

Western Electrician (Chicago, IL). 'Edison's Visit to Chicago.' 23 May 1891, 295.

Wheeling Daily Intelligencer (Wheeling, WV). 'Untitled' (Edison's promised Kinetograph). 25 May 1891.

New York Sun. 'The Kinetograph: Edison's latest and most surprising device.' 28 May 1891.

Asheville Daily Citizen (Asheville, NC). 'Edison's Latest.' 28 May 1891.

The Evening World (New York). 'The Kinetograph.' 28 May 1891.

Times (London). 'Mr. Edison's Latest Invention.' 29 May 1891.

New York Times. 'A Move in the Right Direction – Mr. Edison's Latest.' 29 May 1891.

Pall Mall Gazette (London). 'Occasional Notes.' 29 May 1891.

Pall Mall Gazette (London). 'Mr. Edison's Latest Invention.' 29 May 1891.

Indianapolis Herald (Indianapolis, IN). 'Light Wedded to Sound.' 31 May 1891.

English Mechanic. 'Edison's Kinetograph' (reprinted in the *Times* London). 5 June 1891, 310.

Engineering. 'Edison's Kinetograph.' 5 June 1891, 678.

American Engineer. 'Reflection on Western Journalism.' 6 June 1891, 223.

Punch. 'What it May Come to in London.' 6 June 1891.

The Newcastle Weekly Courant. 'Science and Invention.' 13 June 1891.

Lathrop, George Parsons. 'Edison's Kinetograph.' *Harpers Weekly.* 13 June 1891, 446-447.

Electrical World. 'The Kinetograph.' 13 June 1891, 431.

Scott, George B., 'Edison's Latest Marvel: The Kinetograph,' letter to the editor. *English Mechanic.* 19 June 1891, 358-359.

Bickle, Jebus. 'Edison's Kinetograph,' letter to the editor, *English Mechanic,* 19 June 1891, 359-360.

The Boys Comic Journal (London). 'Fickle Minded.' 20 June 1891, 320.

Scientific American. 'Edison's Kinetograph.' 20 June 1891, 393.

Illustrated American. 'Punch and the Kinetoscope.' 20 June 1891, 224.

Illustrated American. 'Mr. Edison's Kinetograph.' 20 June 1891, 194.

The Electrical Engineer. 'The Edison Kinetograph.' 24 June 1891, 708.

Manufacturer and Builder. 'Edison's Kinetograph.' June 1891, 138.

American Journal of Photography. 'Edison's Latest' (reprinted from the *Philadelphia Daily News*). June 1891, 338.

Rudge, W. A. 'The Kinetograph and Other Optical Apparatus,' letter to the editor, *English Mechanic.* 3 July 1891, 406.

Eldridge, E. letter to the editor, *English Mechanic.* 3 July 1891, 406.

Constant Buyer, letter to the editor, *English Mechanic.* 3 July 1891, 406.

Dor, Nun. 'The Kinetograph,' letter to the editor, *English Mechanic.* 10 July 1891, 431.

Life. 'We May Be Disappointed.' 2 July 1891, 412.

Ogle, Jabez, 'The Kinetograph,' letter to the editor, *English Mechanic.* 24 July 1891, 479.

Davis, Oscar K. 'Edison in his Laboratory.' *Electricity.* 22 July 1891, 5.

Eldridge, H. letter to the editor (reply to Nun Dor), *English Mechanic.* 24 July 1891, 479.

Chambers' Journal. 'The Month – Science and Arts.' 25 July 1891, 477.

T.R. 'Seeing by Electricity,' letter to the editor, *English Mechanic.* 31 July 1891, 504-505.

Leisure Hour. 'Edison's Latest Invention: The Kinetograph.' August 1891, 711-712.

August, E. 'Seeing by Electricity,' letter to the editor (Reply to T. R.), *English Mechanic.* 28 August 1891.

Cassell's Family Magazine. 'The Kinetograph.' August 1891, 575-576.

Carpenter, Frank. 'Edison in his Den.' *The Pittsburg Dispatch.* 5 November 1891.

Punch. 'Telephonic Theater-Goers (a Sketch at the Electrical Exhibition).' 30 April 1892, 208.

Harper's Weekly. 'Electricity at the Fair.' 16 July 1892.

Phonogram. 'The Kinetograph.' October 1892, 217-219.

Scientific American. 'First Public Exhibition of Edison's Kinetograph.' 20 May 1893.

The Photographic News. 'Edison's Kinetograph.' 16 June 1893.

Phillips, Barnet. 'The Record of a Sneeze.' *Harper's Weekly.* 24 March 1894.

Trewman's Exeter Flying Post or Plymouth and Cornish Advertiser (UK). 'Local Gossip.' 13 May 1893.

Scientific American. 'First Public Exhibition of Edison's Kinetograph.' 20 May 1893.

The Review of Reviews. 'The Kinetograph.' May 1894.

The Electrical World. 'The Kineto-phonograph.' 16 June 1894.

Edison, Thomas. 'Edison's Invention of the Kineto-Phonograph: Introduction.' *Century Magazine.* June 1894.

Dickson, W. K. L., and Antonia Dickson. 'Edison's Invention of the Kineto-Phonograph.' *Century Magazine.* June 1894.

The Literary Digest. 'The Kineto-phonograph.' 21 July 1894.

The Electrical Engineer. 'The Edison Kinetoscope.' 7 November 1894.

Dickson, W. K. L., and Antonia Dickson. *The life and inventions of Thomas Alva Edison.* New York: TY Crowell, 1894.

New York Times. 'An Authentic Life of Edison,' review of *The Life and Inventions of Thomas Alva Edison*, by W. K. L. and Antonia Dickson. 11 November 1894.

The Dial. 'The Life and Work of Edison,' review of *The Life and Inventions of Thomas Alva Edison*, by W. K. L. and Antonia Dickson. 16 November 1894.

The Electrical Engineer. Review of *The Life and Inventions of Thomas Alva Edison*, by W. K. L. and Antonia Dickson. 28 November 1894, 443.

The Saturday Review of Politics, Literature, Science and Art. 'The Electric Barnum,' review of *The Life and Times of Thomas Alva Edison*, by W. K. L. and Antonia Dickson. 1 December 1894, 601-602.

Western Electrician. 'A Scotch Estimate of Edison.' 5 January 1895.

Lathrop, George Parsons. 'If Edison Can Find the Way.' *The Youth's Companion*. 10 January 1895.

The Electrical World. Review of *The Life and Inventions of Thomas Alva Edison*, by W. K. L. and Antonia Dickson. 23 February 1895.

Popular Science. Review of *The Life and Inventions of Thomas Alva Edison*, by W. K. L. and Antonia Dickson. February 1895.

Dickson, Antonia. 'Wonders of the Kinetoscope.' *Frank Leslie's Illustrated Magazine*. March 1895.

Tyrell, Henry. 'Edison,' *Frank Leslie's Popular Monthly*. March 1895.

Engineering and Mining Journal. Review of *The Life and Inventions of Thomas Alva Edison*, by W. K. L. and Antonia Dickson. 20 July 1895.

Matthews, Brander. 'The Kinetoscope of Time.' *Scribner's*. December 1895, 733-745.

Dickson, W. K. L. and Antonia Dickson. *History of the Kinetograph, Kinetoscope, and Kineto-phonograph*. New York: Albert Bunn, 1895; New York, Museum of Modern Art, 2000.

To-Day: A Weekly Magazine-Journal. 'Mr. Maskelyne and the Animated Photograph Craze.' 16 May 1896.

Mundell, Frank. *The Story of Edison and the Wonders of Electricity*. London: Jarrod & Sons, 1897.

The Spectator. Review of *The Story of Edison and the Wonders of Electricity*, by Frank Mundell. 20 November 1897.

The Saturday Review. 'The Electrical Barnum Again.' 20 November 1897, 560.

Punch. 'Our Booking Office,' review of *The Story of Edison*, by Frank Mundell. 27 November 1897.

Brief for Complainant, Thomas A Edison v. The American Mutoscope Company and Benjamin F. Keith, 1898 (TAED QM001195).

Litigation Records Series, Thomas A. Edison v. American Mutoscope Company, 1898-1900 (TAED QM001).

Seeing by Electricity

Von Siemens, Werner. 'Action of Light on Selenium.' *Nature* 13, no. 334 (1876): 407.
Scientific American. 'Siemens' Sensitive Electric Eye.' 8 December 1876, 374.
Figuier, Louis. 'Le Telectroscope,' in *L'Annee Scientifique et Industrielle* (1877).
The Electrician. 'The Electroscope,' letter to the editor, *New York Sun*. 30 March 1877.
New York Sun. 'Untitled' (The Electrotherm). 1 April 1877.
J. F. W. letter to the editor, *Nature* 18, no. 450 (13 June 1878): 169.
De Paiva, Adriano. *Telephony, Telegraphy, and Telescopy*. Ciombra, Portugal, 1878.
Le Mondes. 'Telectroscope.' 16 January 1879.
Times (London). 'The Telectroscope' (Senlecq). 27 January 1879.
'Notes' (The Telectroscope). *Nature* 19, no. 482 (January 1879): 278.
English Mechanic. 'The Telectroscope.' 31 January 1879, 509.
The Electrician. 'The Telectroscope.' 1 February 1879.
Redmond, Denis D. 'An Electric Telescope,' letter to the editor, *English Mechanic*. 7 February 1879, 540.
Telegraphic Journal and Electrical Review. 'The Telectroscope.' 15 February 1879.
Morshead, W. letter to the editor, *English Mechanic*. 21 February 1879, 586.
Glew, F. H. 'Electric Telescope,' letter to the editor, *English Mechanic*. 21 February 1879, 586.
Scientific American. 'A Novel and Curious Instrument. The Telectroscope.' 8 March 1879.
The Manufacturer and Builder. 'The Telectroscope Used for Photographing Over Telegraph Wires.' April 1879, 86.
Scientific American. 'The Telectroscope.' 17 May 1879.
Harper's New Monthly Magazine. 'Editor's Scientific Record: Miscellaneous.' May 1879, 947.
McClure, J. B. *Edison and His Inventions*. Chicago, IL: Rhodes & McClure Publishing Company, 1879.
Daily Times (Bethlehem, PA). 'Untitled' (Licks' Diaphote). 10 February 1880.
American Manufacturer and Iron World. 'The Firm of Connelly Bros. & McTighe' (Telephote). 20 February 1880.
New York Times. 'Dr. H.E. Licks, of South Bethlehem, Penn.' 25 February 1880.
New York Times. Editorial (Diaphote and Telephote), 2 March 1880.
'Physical Notes.' *Nature* 21, no. 545 (April 1880): 575.
Ayrton, W. E. and John Perry. 'Seeing by Electricity,' letter to the editor, *Nature* 21, no. 546 (April 1880): 589.
Middleton, H. 'Seeing by telegraph,' letter to the editor, *Times* (London). 24 April 1880.
Gordon, J. E. H. 'Seeing by electricity,' letter to the editor, *Nature* 21, no. 547 (April 1880): 610.

English Mechanic. 'Seeing by Telegraph.' 30 April 1880, 177-178.

Reynolds Newspaper (London). 'Scraps from the Comic Journals.' 9 May 1880.

Ayrton, W. E., and John Perry. 'Seeing by Telegraph' (reply to Gordon), letter to the editor, *Nature* 22, no. 550 (May 1880): 31.

Bolton, H. E. 'Seeing by electricity,' letter to the editor, *English Mechanic.* 14 May 1880, 235.

Scientific American. 'Seeing by Electricity' (Carey). 5 June 1880, 355.

Sawyer, W. E., letter to the editor (reply to 'Seeing by Electricity,' Carey), *Scientific American.* 12 June 1880, 373.

The Electrician. 'Report.' 19 June 1880, 52.

Manufacturer and Builder. 'Seeing by Electricity.' 7 July 1880, 158.

New York Times. 'The Photophone.' 30 August 1880.

Thompson, Silvanus. 'The Photophone.' *Nature* 22, no. 569 (September 1880): 481.

Bell, A. Graham. 'Selenium and the Photophone' (Lecture delivered at the Boston meeting of the American Association). *Nature* 22, no. 569 (September 1880): 500.

'Bell's Photophone.' *Nature* 23, no. 575 (November 1880): 15.

Bidwell, Shelford. 'The Photophone.' *Nature* 23, no. 577 (November 1880): 58.

Molison, A.R. letter to the editor, *Nature* 23, no. 578 (November 1880): 78.

Scribner's Monthly. 'The Photophone.' November 1880, 164-165.

Independent Statesman (Concord, NH). 'Multiple News Items' (Seeing by Electricity). 6 December 1880.

Scientific American. 'Bell's Photophone.' 1 January 1881, 1, 4.

Senlecq, Constantin. 'The Telectroscope.' *The Electrician.* 5 February 1881, 141-142.

Bidwell, Shelford. 'Telephotography.' *Nature* 23 no. 589 (February 1881): 344.

English Mechanic. 'The Telectroscope.' 11 February 1881.

Ayrton, W. E., and John Perry. 'Seeing by Electricity' (Paper communicated to the Physical Society). *Nature* 2,3 no. 592 (March 1881): 423.

Scientific American Supplement. 'The Telectroscope.' 9 April 1881, 4382.

London Standard. 'Future of Electric Invention.' 26 March 1881.

New York Times. 'Future of Electric Invention.' 16 April 1881.

Bidwell, Shelford. 'Telegraphic photography.' *Journal of the Society of Telegraph Engineers and Electricians* 10, no. 38 (September 1881): 354-360.

Lucas, William. 'The Telectroscope, or "Seeing by Electricity",' letter to the editor, *English Mechanic.* 21 April 1882, 151-152.

Seward, Alfred. 'Seeing by Electricity' (response to Lucas), letter to the editor, *English Mechanic.* 28 April 1882, 177.

Atkinson, L.B. (L.I.B.A.) 'The Telectroscope—Dynamomahcines—Heat Equivalent' (response to Lucas), letter to the editor, *English Mechanic.* 5 May 1882, 194.

Robida, Albert. *Le Vingtième Siècle* (*The Twentieth Century*). Paris: Librairie Illustree, 1882; Middletown, CT: Wesleyan University Press, 2004.

The Electrician. 'Remarkable Electrical Discovery.' 31 March 1883, 460.

The Electrical Engineer (UK). 'The British Association' (Announcing progress in 'electric photography at a distance'). 23 August 1889, 150-151.

Telegraphic journal and Electrical Review. 'Seeing to a distance by electricity.' 29 November 1889, 609-611.

Atikinson, L.B. 'Seeing to a distance by electricity.' *Telegraphic Journal and Electrical Review*. 13 December 1889, 683.

The Electrician. 'Seeing by electricity.' 7 March 1890, 448-450.

The Electrical Engineer. 'Seeing by Electricity' (Hughes and Preece). 9 July 1890, 52.

The Electrical World. 'Seeing by Electricity.' 12 July 1890, 26-27.

The Electrical World. 'Miscellaneous Notes' (English Electricians Have a Joke) (Hughes and Preece). 12 July 1890, 32.

The Electrical World. 'A Remarkably Successful Joke' (Hughes and Preece). 9 August 1890, 32.

Sutton, Henry. 'Tele-Photography' (The Telephane). *Telegraphic Journal and Electrical Review*. 6 November 1890, 549-551.

'Reproduction of Light by Electricity' (reprinted from *Iron*). *American Journal of Photography*. 1 October 1890, 290.

T.R. 'Seeing by electricity,' letter to the editor, *English Mechanic*. 26 June 1891, 387-388.

August, E. 'Seeing by Electricity,' letter to the editor, *English Mechanic*. 3 July 1891, 409.

T.R. letter to the editor, *English Mechanic*. 31 July 1891, 503-4.

Le Pontois, Leon. 'The Telectroscope.' *Electricity*. 10 May 1893, 226.

Le Pontois, Leon. 'The Telectroscope.' *The Electrical Engineer*. 17 May 1893, 485.

Jenkins, C. Francis. 'Transmitting Pictures by Electricity.' *The Electrical Engineer*. 25 July 1894.

Electrical Engineer. 'Gooseberries in March.' 4 March 1898, 257.

Electrical Engineer. 'Next, Please!' 11 March 1898, 304-5.

Neue Weiner Tageblatt (Vienna, Austria). 'Der Elektrische Ferneher, besuch bei Hern Ein Jan Szczepanik.' 17 March 1898.

Boston Daily Advertiser. 'The Far-Seer.' 21 March 1898.

New York Tribune. 'Herr Szczepanik's Telectroscope.' 23 March 1898.

Electrical Engineer. 'Telectroscopy.' 25 March 1898, 354.

Horowitz, Johannes. 'That New Telectroscope.' *New York Times*. 3 April 1898.

Electrical Engineer. 'The Fernseher Again.' 15 April 1898, 449.

Electrical Engineer. 'More Szczepanik.' 22 April 1898, 483.

Illustrated London News. 'The Latest Triumph of Electricity.' 23 April 1898.

The Milwaukee Journal. '"Genius" Triumph over Great Obstacles.' 23 April 1898.

Weekly Rocky Mountain News. 'Science and Discovery.' 5 May 1898.

Weekly Rocky Mountain News. 'Inventor of the Telectroscope.' 12 May 1898.

Oppenheim, William George. letter to the editor, *American Amateur Photographer.* 20 May 1898, 276.

New York Sun. 'Cry for a Telectroscope.' 22 May 1898.

Milwaukee Daily Journal. 'Personal and Impersonal.' 24 May 1889.

Lanterna Magica (Leipsig). 'Das Telelectroscop.' May 1898.

The Bristol Mercury and Daily Post. 'A Great Invention.' 2 June 1898.

Electrical Engineer. 'More Szczepanik.' 3 June 1898, 675.

Electrical Engineer. 'The German Press and Szczepanik.' 1 July 1898, 3.

Scientific American supplement. 'Untitled' (Dussaud's Teleoscope). 2 July 1898, 1893.

Armengaud, M. 'The Dussaud Teleoscope' (reprinted from *La Nature*). *Scientific American Supplement.* 2 July 1898, 18793.

American Monthly Review of Reviews. 'The Telelectroscope and its Inventor.' 18 July 1898, 93-4.

Electrical Engineer. 'More Szczepanik.' 29 July 1898, 129.

Twain, Mark. 'The Austrian Edison Keeping School Again.' *The Century.* August 1898, 630-631.

Twain, Mark. 'From the "London Times" of 1904.' *The Century.* November 1898, 100-105.

Moffett, Cleveland. 'Seeing by Electricity.' *Pearson's Magazine.* October 1899, 490.

Glasgow Herald. 'Paris Exhibition.' 12 April 1900.

Cook, Charles Emerson. 'Pictures By Telegraph.' *Pearson's Magazine.* April 1900, 345-348.

The Electrician. 'The International Electricity Congress.' 21 September 1900.

Kinsman, F.E. 'Seeing by Electricity.' *New York Times.* 10 March 1901.

Perskyi, Constantin. 'Television Using Electricity.' In *Congres International D. Electricite (International Congress on Electricity of 1900).* Paris: Gauthier –Villars, 1901.

Electrical Review. 'A Talk with Edison.' 10 January 1903, 33.

Lux, Fritz. *Der Elektrische Fernseher.* Ludwigshafen, 1903.

Maver, William, Jr. 'Seeing by Electricity' (Fowler and Thompson Televue). *Cassier's Magazine.* October 1906, 519-520.

The Electrical Age. 'Seeing by Electricity.' November 1, 1906, 374.

Washington Post. 'Seeing by Electricity' (Fowler and Thompson Televue). November 11, 1906.

The Illuminating Engineer. 'Seeing by Electricity' (Fowler and Thomspson Televue). 1 April 1907, 149.

Los Angeles Times. 'Seeing by Electricity.' 11 August 1912.

Cassier's Engineering Monthly. 'Action at a Distance.' August 1912, 164.

Scientific American Supplement. 'Action at a Distance.' 17 January 1914, 39.

Manchester Guardian (UK). 'Motion and the Motor-Car: Seeing by Electricity.' 28 May 1914.

Machine-Age Television

New York Times. 'The Latest Scientific Marvel: An Electric Eye.' 23 December 1906.

Bidwell, Shelford. 'Telegraphic Photography and Electric Vision.' *Nature* 78, no. 2014 (June 1908): 105.

Campbell-Swinton, A. A. 'Distant Electric Vision,' letter to the editor, *Nature* 78, no. 2016 (June 1908): 151.

Grimshaw, Robert. 'The "Telegraphic Eye".' *Scientific American.* 1 April 1911.

Scientific American Supplement. 'Prof. Rozing's "Electric Eye" – A New Apparatus for Television.' June 1911.

Campbell-Swinton, A. A. 'Distant Electric Vision.' *Times.* 15 November 1911.

Campbell-Swinton, A. A. 'Presidential Address (with concluding remarks by Silvanus Thompson).' *Journal of the Röntgen Society.* January 1912, 1-15.

St. Louis Dispatch. 'Persons Miles Apart See Each Other by Phone.' 29 May 1914.

Secor, H.W. 'Television, Of the Projection of Pictures Over a Wire.' *Electrical Experimenter.* August 1915, 131-132, 172-174.

Johnson, V.E. *Modern Inventions.* New York: FA Stokes, 1915.

Gernsback, Hugo. 'Television and the Telephot.' *Electrical Experimenter.* May 1918.

Martin, Marcus. 'Television.' *Electricity.* 21 October 1921, 577.

Campbell-Swinton, A. A. 'The Possibilities of Television with wire and wireless.' *Wireless World.* 9 April 1924; 16 April 1924; 23 April 1924.

Hopkins, A. A. 'A Machine that is More than Human.' *Scientific American.* December 1925, 336-337.

Campbell-Swinton, A. A. 'Electric Television.' *Nature* 118, no. 2973 (October 1926): 590.

Dinsdale, Alfred. *Television: Seeing by Wireless.* London: Sir I. Pitman & Sons, 1926.

New York Times. 'Glasgow Listens to Sound of Faces.' 4 February 1927.

Armagnac, Alden. 'We'll Soon SEE by Radio, Too!' *Popular Science Monthly,* March 1927, 37.

Indianapolis Star. 'Science's Latest Marvel.' 9 April 1927.

Life. 'To a Waiting World.' 28 April 1927, 7.

Mayer, Lloyd. 'Just Between Us Girls.' *Life.* 5 May 1927, 14.

Ryan, Phyllis. 'Television.' *Life.* 5 May 1927, 18.

Secor, H. W. 'Television Perfected at Last.' *Science and Invention.* June 1927, 108.

Popular Science Monthly. 'Television—At Last!' June 1927, 11-13, 130.

Albelli, A. A. 'What's Next in Television?' *Popular Mechanics.* July 1927, 1-3.

Down, George Lee, Jr. 'Next We'll See to Paris.' *Popular Science.* September 1927, 23.

Baker, Thomas. *Wireless Pictures and Television.* New York : Van Nostrand, 1927.

New York Times. 'We Catch a Glimpse of Tomorrow.' 16 January 1928.

The Independent. 'When the Movies Fly Away.' 25 February 1928, 172.

Television. 'Television 1873-1927: A brief outline of what has been accomplished in little over a century.' March 1928, 10-11.

New York Times. 'Radio Board Tests Television Process.' 28 April 1928.

Armagnac, Alden. 'Television Brought into the Home.' *Popular Science Monthly.* April 1928, 20-21, 143.

Van Zile, Edward. 'New Perils of Television.' *The Spur.* 1 May 1928, 78.

Smith, Agnes. 'The Radio: Television (In Case You're Interested).' *Life.* 23 August 1928, 7.

Lane, Alfred. 'The Real Facts About Television.' *Popular Science Monthly,* September 1928, 43-44.

The Independent. 'Remarkable Remarks.' 13 October 1928, 2.

Sarnoff, David. 'Forging an Electric Eye to Scan the World.' *New York Times.* 18 November 1928.

Popular Mechanics. 'What Television Offers You.' November 1928, 820-824.

Larner, Edgar T. *Practical Television.* New York: D. Van Nostrand, 1928.

Stuart, Arthur A. 'Strange eyes that never sleep.' *Popular Science.* May 1929, 51.

New York Times. 'Radio's Flickering 'Eyes' Now Sensitive to Color.' 7 July 1929, 124.

Miller, James. 'The Latest in Television.' *Popular Mechanics.* September 1929, 472-474.

Armagnac, Alden. 'Now—Television in Natural Colors.' *Popular Science Monthly.* September 1929, 25.

Projection Engineering. 'The Photoelectric Cell: Light Energy and the Photoelectric Effect.' November 1929, 35.

Moseley, Sydney. *Television Today and Tomorrow.* London: Isaac Pitman, 1929.

Sheldon, H.H. and Edgar Norman Grisewood. *Television: Present Methods of Picture Transmission.* New York: D. Van Nostrand Company, Inc., 1929.

Yates, Raymond F. *ABC of Television; Or, seeing by Radio.* New York: Norman W. Henley Publishing, 1929.

Jenkins, C. F. *Radiomovies, Radiovision, Television.* Washington, D.C.: National Capital Press, inc., 1929.

Modern Mechanics. 'Behind the Scenes with Television.' January 1930, 100-101.

Popular Science. 'Electric Eye that Never Sleeps' (Westinghouse Ad). February 1930, 7.

Nason, C.H.W. 'Photocell Applications in Projection.' *Projection Engineering.* April 1930, 15-17.

Dunlap, Orrin E. 'Television Brought Nearer the Home.' *New York Times.* 25 May 1930.

Jenkins, CF. 'Life Size Radio Movies are Coming.' *Modern Mechanics and Inventions.* May 1930, 70-73.

Duffus, R.L. 'The Magic of a Vast Laboratory.' *New York Times.* 8 June 1930.

General Electric News. 'Television Makes Good on the Stage.' 20 June 1930.

Cockaday, Laurence. 'The Latest Developments in Television Methods.' *Radio Craft.* July 1930, 22-4.

Caldwell, O.H. 'The Tiny Electron Works Wonders for Man.' *New York Times.* 6 July 1930.

Popular Mechanics. 'The First Television Show.' August 1930, 177-179.

Martin, Robert. 'Mystery Cell Aids Television.' *Modern Mechanics.* August 1930, 15-17.

Modern Mechanics. 'Television Now Gives Radio Eyes and Ears.' August 1930. 168-171.

First Demonstration of Two Way Television: A Development of the Bell System. New York: Bell Labs, 1930.

New York Times. 'Radio-Images are Clarified.' 11 January 1931.

Goldsmith, Alfred. 'Electrical Entertainment: A Glimpse of the Future.' *New York Times.* 22 March 1931.

Nason, C.H.W. 'A Short Course in Television.' *Television News.* March-April 1931, 30-31, 71.

Miller, J. Earle. 'What's Keeping Television Out of Your Home?' *Modern Mechanics.* May 1931, 98.

Waltz, George H. 'Get it on Television.' *Popular Science.* July 1931, 16-17, 136.

'Results of $50.00 'New Word' Contest.' *Television News.* July 1931, 211.

Olpin, A. R. 'New Types of Photoelectric Cells.' *Television News.* July 1931, 212.

Tichenor, George. 'Television: Is It Getting Anywhere?' *Forum and Century.* October 1931, 218.

Everyday Science and Mechanics. 'Latest Television Broadcast Station.' November 1931, 690-691, 721.

Felix, Edgar H. *Television: Its Methods and Uses.* New York: McGraw-Hill, 1931.

Dunlap, Orrin. *Outlook for Television.* New York: Harper and Brothers, 1932.

Dinsdale, Alfred. *First Principles of Television.* New York: Wiley, 1932.

Sarnoff, David. 'Where Television Stands Today.' *Modern Mechanics.* April 1932, 40-46, 170.

Caldwell, Orestes. 'Tireless Electric Eye Far Outdoes the Human.' *New York Times.* 16 July 1933.

Martin, Robert. 'Electric Eye Guides Ghostly Hands at a Thousand Jobs.' *Popular Science.* August 1933, 20-21.

Peck, A.P. 'A Real "Electric Eye".' *Scientific American*, September 1933, 117.

Popular Mechanics. '"Mosaic" Television for the Home.' September 1933, 321-324.

Armagnac, Alden. '"Human eye" Camera Opens New Ways to Television.' *Popular Science Monthly.* September 1933, 11-13.

Bell, D'Orsay. 'Alternatives to Home Television.' *Wireless World.* 18 January 1935, 72.

Popular Mechanics. 'Electronic Camera "Shoots" Television Images.' June 1935, 878.

Popular Mechanics. 'London Station to Serve Ten Million.' June 1935, 879.

Fawcett, Arthur. *Electric Eyes: A Concise and Elementary Description of the Photo-electric Cell, for the Non-technical Reader: Its Uses in Industry, and Its Uses and Short-comings.* Salisbury: The Tintometer, 1954

The Bell Labs Ikonophone (1927-1930)

'Preliminary Demonstration Program,' for 7 April 1927 Television demonstration, Container 8, Frederic Eugene Ives and Herbert Eugene Ives Papers, Manuscript Division, Library of Congress, Washington, D.C.

'Program of Demonstration,' for First Television Demonstration (April 7, 1927), Container 3, Frederic Eugene Ives and Herbert Eugene Ives Papers, Manuscript Division, Library of Congress, Washington, D.C.

New York Times. 'Far-off Speakers Seen as Well as Heard Here in a Test of Television: Like a Photo Come to Life.' 8 April 1927.

Troy Record. 'Television Now Reality; Device Demonstrated.' 8 April 1927.

New York Times. 'Washington Hails the Test: Operator There Puts Through the Calls as Scientists Watch.' 8 April 1927.

'Television.' Special Issue, *Bell Laboratory Record* 4 no. 3 (May 1927). Articles including: 'The Demonstration at Washington,' 'Television by Radio,' 'Research and Development Leading to Television,' 'Physical Principles and Apparatus,' and 'Television—A Group Achievement.'

Ives, Herbert. 'Television.' *Bell System Technical Journal* (October 1927): 551-559.

Gray, Frank, J.W. Horton, and R.C. Mathes. 'The Production and Utilization of Television Signals.' *Bell System Technical Journal* (October 1927): 560-577.

Television: An Achievement in Electrical Communication. New York: AT&T/ Bell Telephone Laboratories, November 1927.

John Mills. *Through Electrical Eyes: An Elementary Exposition of the Physics and Chemistry Involved in Television.* New York: AT&T/Bell Telephone Laboratories, March 1928.

New York Times. 'Television Shows Panoramic Scene Carried by Sunlight.' 13 July 1928.

Ives, Herbert. 'Television in Colors.' *Bell Laboratories Record.* July 1929.

Ives, Herbert, and A.J. Johnsrud. 'Television in Colors by a Scanning Beam Method.' *Journal of the Optical Society of America* 20 (January 1930): 11-22

Washington Post. 'Two-Way Television Closes Two-Mile Gap.' 9 April 1930.

New York Times. 'Two-Way Television in Phoning Tested.' 10 April 1930.

Kennedy, T.R. 'Speakers on Phone See Images of Each Other.' *New York Times.* 13 April 1930.

New York Times. 'Ethereal Images Baffle Experts.' 20 April 1930.

Phillips, H.I. 'The Once Over: The Television Telephone.' *Boston Globe.* 19 April 1930.

'Two-Way Television.' *The Scientific Monthly* 30 (May 1930): 476-480.

Duffus, R.L. 'The Magic of a Vast Laboratory.' *New York Times.* 8 June 1930.

Brown, Art. 'Television is Ready for Business.' *Nation's Business,* June 1930, 47.

Ives, Herbert E., Frank Gray, and M. W. Baldwin. 'Image Transmission System for Two-Way Television.' *Bell System Technical Journal* 9 no. 3 (June 1930): 448-469.

New York Times. 'Deaf "Hear" 2 Miles in Television Test.' 3 July 1930.

Popular Science. 'Talk, Hear, SEE on This Phone: Two-Way Television Is Demonstrated in Laboratory as an Engineering Stunt.' July 1930, 22, 123.

New York Times. '100 Trade Leaders Tour Laboratories.' 8 October 1930.

Two-Way Television and A Pictorial account of its background. New York: AT&T/ Bell Laboratories, 1930.

'What Hath God Wrought!' Newspaper clipping, February 18, 1931. Container 4, Frederic Eugene Ives and Herbert Eugene Ives Papers, Manuscript Division, Library of Congress, Washington, D.C.

Ives, Herbert. 'Some Optical Features in Two-Way Television.' *Bell System Technical Journal* 10 (February 1931): 265-271.

'Demonstration of Television' invitation card (April 13, 1927), Container 3, Frederic Eugene Ives and Herbert Eugene Ives Papers, Manuscript Division, Library of Congress, Washington, D.C.

Invitation card, Two-way television demonstration, 1930, Container 3, Frederic Eugene Ives and Herbert Eugene Ives Papers, Manuscript Division, Library of Congress, Washington, D.C.

Goldsmith, Alfred. 'Electrical Entertainment: A Glimpse of the Future.' *New York Times*. 22 March 1931.

Miller, Earle. 'What's Keeping Television Out of Your Home?' *Modern Mechanics*. May 1931, 98-102, 198, 200.

Popular Mechanics. 'Two-Way Television Improvements Are Sought.' May 1931, 785.

Ives, Herbert. 'Television: 20th Anniversary.' *Bell Laboratories Record* 25 no. 5 (May 1947): 190-193.

The Reporter (Bell Telephone Laboratories Magazine). 'Birth of Television: TV Marks its Birthday.' April 1957, 4-6, 14.

New York Times. 'Dr. Herbert Ives, Pioneer in TV, Dies.' 15 November 1953.

Montclair Times. 'H. E. Ives, Physicist, Succumbs.' 19 November 1953.

The Reporter (Bell Telephone Laboratories Magazine). 'In Memoriam.' December 1953, 23. Container 3, Frederic Eugene Ives and Herbert Eugene Ives Papers, Manuscript Division, Library of Congress, Washington, D.C.

Physics Today. Necrology, January 1954. Container 3, Frederic Eugene Ives and Herbert Eugene Ives Papers, Manuscript Division, Library of Congress, Washington, D.C.

'Herbert E. Ives.' *Biographical Directory of American Men in Science*, Part I: Physical Science, Ninth Edition. Container 3, Frederic Eugene Ives and Herbert Eugene Ives Papers, Manuscript Division, Library of Congress, Washington, D.C.

Buckley, Oliver, and Karl K. Darrow. 'Herbert Eugene Ives (1882-1953).' Washington, DC: National Academy of Sciences, 1956.

Illuminating Engineering

Induction Letter, Illuminating Engineering Society, September 25, 1908. Container 9, Frederic Eugene Ives and Herbert Eugene Ives Papers, Manuscript Division, Library of Congress, Washington, D.C.

Ives, Herbert. 'Daylight Efficiency of Artificial Light Sources.' *The American Physical Society* 28 (1909): 458.

Electrical Review and Western Electrician. 'Physical Laboratory of the National Electric Lamp Association: A Laboratory for Research in Illumination and Electric Lighting' (reprinted in *Science*, December 24, 1909). 10 September 1910, 507-511.

The Illuminating Engineer. 'The Johns Hopkins University Course of Lectures on Illuminating Engineering.' July 1910, 256-257.

Codman, John. 'Advent of Illuminating Engineering.' In *Technology and Industrial Efficiency*, edited by the Congress of Technology, 277-278. New York: McGraw-Hill, 1911.

Good Lighting and Illuminating Engineer. 'A Year's Progress in Illuminating Engineering.' January 1912, 2.

Light: Its Use and Misuse. New York: Illuminating Engineering Society, 1912.

Hirshberg, Leonard Keene. 'Why Our Eyes Very Often Deceive Us.' Newspaper clipping, *Evening Sun*, 20 July 1914, Container 4, Frederic Eugene Ives and Herbert Eugene Ives Papers, Manuscript Division, Library of Congress, Washington, D.C.

Ives, Herbert. 'An Illuminating Engineer's Conception of an Ideal Light.' *Transactions of the American Electrochemical Society* 17 (1915): 419-433.

Ives, Herbert. 'Simple Methods of Spot-lighting in the Home.' *Lighting Journal* (June 1916): 121-122.

Ives, Herbert. 'A Day and Night Window.' *Lighting Journal* (August 1916): 172-173

Ives, Herbert. 'Evidence that Persistence of Vision is a Physical Conduction Phenomenon.' *Journal of the Franklin Institute* (June 1917): 779.

Ives, Herbert. 'Proposed Standard Method of Colorimetry.' *Journal of the Optical Society of America* (November 1921): 469-478.

Cady, Francis, and Henry Baldwin Dates, eds. *Illuminating Engineering.* New York: Wiley, 1925.

Luckiesh, Matthew, and Frank Moss. 'The New Science of Seeing.' *Transactions of the Illuminating Engineering Society* 25 (January 1930): 15-39.

Ives, Herbert, and E. F. Kingsbury. 'The Applicability of Photoelectric Cells to Colorimetry.' *Journal of the Optical Society of America* 21 (September 1931): 541-563.

Jolley, Leonard, John Waldram, and George Wilson. *The Theory and Design of Illuminating Engineering Equipment.* New York: Wiley, 1931.

Luckiesh, Matthew. 'The Human Seeing-Machine.' *Transactions of the Illuminating Engineering Society* 27 (1932): 699-722.

Judd, Deane. 'The 1931 I. C. E.: Standard Observer and Coordinate System for Colorimetry.' *Journal of the Optical Society of America* 23 no. 10 (1933): 359-373.

Luckiesh, Matthew, and Frank Moss. 'The Human Seeing-Machine.' *Journal of the Franklin Institute* 215 (June 1933): 629-654.

Luckiesh, Matthew, and Frank Moss. *The New Science of Lighting.* Schenectady, NY: Lighting Research Lab, General Electric, 1934.

Luckiesh, Matthew, and Frank Moss. 'The Motion Picture Screen as a Lighting Problem.' *Journal of the Society of Motion Picture Engineers* 26 (May 1936): 578-591.

Luckiesh, Matthew, and Frank Moss. *The Science of Seeing.* New York: D. Van Nostrand Co, 1937.

Reeb, O. 'A Consideration of the Screen Brightness Problem.' *Journal of the Society of Motion Picture Engineers* 32 (May 1939): 485-494.

Jones, Lloyd. 'The Historical Background and Evolution of the Colorimetry Report of the OSA.' *Optometry and Vision Science* 21 (August 1944): 317-333.

Nelson, J.H. 'Ideal Seeing Conditions: The Study of the Human Visual System as a Basis for Prescribing Lighting.' *British Journal of Industrial Medicine* 2 (April 1945): 224.

IES Lighting Handbook: A Standard Lighting Guide. New York: Illuminating Engineering Society, 1947.

Lohr, Lenox. *Fair Management, the Story of a Century of Progress Exposition: A Guide for Future Fairs.* Chicago, IL: Cuneo Press, 1952.

Harmon, Leon. 'Analogs and Models of the Human Visual System.' *Optometry & Vision Science* 36 (June 1959): 304-312.

Judd, Deane, and Günter Wyszecki. *Color in Business, Science and Industry.* New York: Wiley, 1963.

McCormick, Ernest. *Human Factors Engineering.* Columbus, OH: McGraw-Hill, 1964.

Judd, Deane. 'Fundamental Studies of Color Vision from 1860-1960.' *NAS Symposium* 55 (1966): 1313-1330.

Jameson, Dorothea, and Leo Hurvich, eds., *Visual Psychophysics.* New York: Springer, 1972.

Granrath, Douglas. 'The Role of Human Visual Models in Image Processing.' *Proceedings of the IEEE* 69 (May 1981): 552-561.

Hunter, Richard, and Richard W. Harold. *The Measurement of Appearance.* New York: Wiley, 1987.

Nemeth, Christopher. *Human Factors Methods for Design: Making Systems Human-centered.* Boca Raton, FL: CRC Press, 2004.

Guastello, Stephen J. *Human Factors Engineering and Ergonomics: A Systems Approach.* Boca Raton, FL: CRC Press, 2013.

Picturephone (1956-1970)

Gould, John. 'Picture, Please!: How you'll be able to see while you talk on the newest version of Mr. Bell's invention.' *New York Times*. 16 September 1956.

Hall, A.D. *A methodology for systems engineering*. New York: Van Nostrand, 1962.

Hall, A.D. 'Experiments with Picturephone Service.' *Bell Labs Record* 42 (April 1964): 114-120.

Goldstein, Edward. 'First Hand: The End of Picturephone.' EEE Oral Histories (2015).

Cutler, C. Chapin. An oral history conducted in 1993 by Andrew Goldstein, (IEEE History Center, Hoboken, NJ).

Schreiber, William. An oral history conducted in 1998 by Frederik Nebeker, (IEEE History Center, Hoboken, NJ).

Pierce, John. An oral history conducted in 1992 by Andy Goldstein, (IEEE History Center, Hoboken, NJ).

Lucky, Robert. An oral history conducted in 1999 by David Hochfelder, (IEEE History Center, Hoboken, NJ).

About the Author

Ivy Roberts is an interdisciplinary film/media-studies scholar who researches in the fields of cultural history, visual culture, and science, technology, and society (STS) Dr. Roberts holds a Ph.D. from Virginia Commonwealth's interdisciplinary programme in Media, Art, and Text. *Visions of Electric Media* has been adapted from her dissertation. Dr. Roberts's next book is an edited volume of critical approaches to Victorian and Radium Age science fiction. She is first and foremost a devoted teacher who holds adjunct faculty positions at Virginia Commonwealth University, the University of Maryland University College, and Southern New Hampshire University.

Index